OXFORD MEDICAL PUBLICATIONS

Methods for the
Economic Evaluation
of Health Care Programmes

4. Were all the important and relevant costs and consequences for each alternative identified?

Even though it may not be possible or necessary to measure and value all of the costs and consequences of the alternatives under comparison, a full identification of the important and relevant ones should be provided. The combination of information contained in the viewpoint statement and programme description should allow judgement of what specific costs and consequences or outcomes it is appropriate to include in the analysis.

An overview of the types of costs and consequences which may be relevant to economic evaluation of health services and programmes is provided in Fig. 3.1, where three categories of costs are shown. Because the costs of a health service or programme are best thought of as the resources it uses up, Category I consists of the costs of organizing and operating the programme. The identification of these costs often amounts to listing the *ingredients* of the programme—both variable costs (such as the time of health professionals or supplies) and fixed or overhead costs (such as light, heat, rent or capital costs). These organizing and operating costs are often referred to as *direct costs* by economists.*

Category II contains costs which are borne by patients and their families. These include any out-of-pocket expenses incurred by patients and/or family members as well as the value of any resources that they contribute to the treatment process. Patients and/or family members sometimes lose time from work while seeking treatment or participating in a health care programme. Such *production losses* are also a cost of the health service or programme in question and are often referred to by economists as *indirect costs* of the service or programme. Care must be taken, however, when including this cost item in an analysis, since its inclusion implies that the cost was incurred as a result of participation in treatment and therefore that the individual's condition was not of a type which would have prevented productive activity anyway.† Finally, the anxiety, and perhaps pain, associated with treatment itself constitute a form of *psychic* cost frequently experienced by patients and their families.

While the above two categories of costs cover most of the cost items relevant to economic evaluations of health services, a third category,

* Health administrators sometimes reserve the term *direct costs* for variable costs only, and may refer to overhead costs as *indirect costs*. In economic evaluations, however, economists employ the term *indirect costs* to denote a quite separate and distinctive type of cost, as explained later in this section. Users of evaluations should be aware of this potential source of confusion.

† The complexity of the relationship between lost work time and the value of forgone output places it beyond the scope and purpose of this chapter. For a discussion of its implications for Category II and III costs see Stoddart (1982).

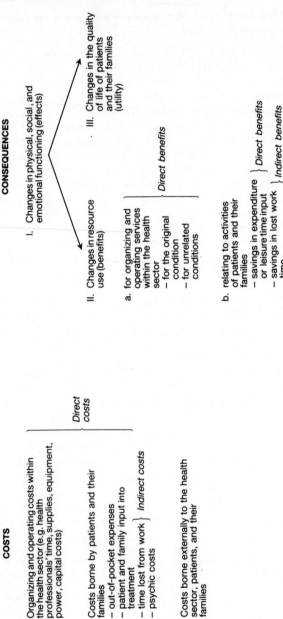

Fig. 3.1. Types of costs and consequences of health services and programmes.

Category III, also warrants mention. Occasionally it may be the case that the operation of a health service or programme changes the resource use in the broader economy outside the health sector. Examples of situations where these factors may be important are:

(a) an occupational health and safety programme (perhaps legislated by government) that changes the production process in an automobile manufacturing plant, thereby using up more resources, perhaps in a more labour-intensive way. These costs are passed on in the increased price of cars and are borne by the purchasers of cars, who are likely not the workers for whom the programme was initiated.

(b) a 55 mile per hour speed limit policy which reduces morbidity and mortality due to accidents, but increases the price of, for example, fruit which now takes longer to arrive (i.e. a higher wage bill for the truck driver?).

In principle these factors should be considered in the economic evaluation, though for many health care programmes they may be insignificant. (Few economic analyses of alternative health care programmes take them into account.)

Three categories of consequences of health services and programmes are also shown in Fig. 3.1. Category I consists of therapeutic outcomes, or effects, of the alternatives in question. Normally, these effects are changes in the physical, social or emotional functioning of individuals. In principle such changes can be measured objectively, and refer only to an individual's ability to function and not to the significance, preference or value attached to this ability by the individual or by others.

The therapeutic effects of a service or programme give rise to two other important categories of consequences. First, the effects may result in changes in resource use in the future (Category II). Within the health sector, less resources may be required for treatment of the condition than otherwise may be the case. For example, an effective hypertension screening programme averts the future cost of caring for stroke victims. The saving in health care utilization attributable to the screening programme is normally referred to by economists as the *direct benefit* of the screening programme. Notice, however, that if we adopt a health system viewpoint, it is sometimes the case that direct benefits are negative, due to the increased utilization of services for different conditions (e.g., arthritis) that patients develop in their newly prolonged lives!

The therapeutic effects of a health service or programme may also affect the resource use of patients and their families. Of particular interest here is the possibility that patients (and family members) may gain

23

increased working time as a result of treatment. These production gains are normally referred to as *indirect benefits* by economists, and their inclusion in economic evaluations is a source of some controversy among analysts.

Three arguments are often put forward against their inclusion. It is sometimes argued (rather narrowly it seems) that health care evaluation should confine itself to changes in resource use within the health sector only, rather than the entire economy. A more serious objection is the assertion that changes in the output of individuals or groups are simply not the grounds upon which we normally make resource-allocation decisions concerning the health of those involved. Therefore it is misleading to enter the value of such changes into a *cost–benefit* calculation. A third criticism is that the valuation of *indirect benefits* (usually through increased earnings of individuals) makes a series of value judgements and assumptions which may only be appropriate in a limited number of cases. While it is not possible to discuss and evaluate these claims here, readers should be aware that the inclusion of *indirect benefits* in a cost–benefit analysis may not be straightforward.* This issue is discussed further in Chapters 5 and 7. [See also Drummond (1981) and Williams (1981).]

The therapeutic effects of health services and programmes also give rise to another extremely important category of consequences, namely changes in the quality of life of patients and their families (Category III). The quality of life produced by the therapeutic effects is distinguished from the effects themselves by the significance or value that patients and their families attach to the effects. It is entirely possible, of course, and in fact likely, that different individuals place differing importance on the same level of physical, social or emotional function. [This was illustrated in Chapter 2 (see p. 13), by the effects of a broken arm on the sign painter and the translator.]

With respect to both the costs and consequences indicated above, it may be unrealistic to expect all relevant items to be measured and valued in the analysis, due to their small size or influence relative to the effort required to measure or value them accurately; however, it is helpful to users to identify as many relevant items as possible. It is particularly

* Those who criticize the inclusion of *indirect benefits*, saying '*You can value a livelihood but you can never value a life!*', appear to be confusing indirect benefits with another type of benefit. This is the *intangible* value we as individuals, and as a society, place on life itself (regardless of earning potential) and on the avoidance of pain and suffering. Although *intangible benefits* (and costs) of health services undoubtedly exist, by their very nature they are difficult to include in a cost–benefit analysis, which expresses costs and consequences in dollars. They presumably are taken into account, however, in *cost–utility* analyses wherein programme effects are translated into a measure of value based on preferences for health states rather than dollars, as discussed in Chapter 6).

important that the outcomes of interest be identified clearly enough for a reader to judge the appropriateness of the type (or types) of economic evaluation chosen. That is, it should be apparent:

— whether a single outcome is of primary interest as opposed to a set of outcomes,
— whether the outcomes are common to both alternatives under comparison, and
— to what degree each programme is successful in achieving each outcome of interest.

Similarly, it is important to know whether the consequences of primary interest are the therapeutic effects themselves (thus implying cost-effectiveness analysis if possible), the net change in resource use (cost–benefit analysis), or the quality of life of patients and their families (cost–utility analysis).

5. Were costs and consequences measured accurately in appropriate physical units?

While identification, measurement, and valuation often occur simultaneously in analyses, it is a good practice for users of evaluation results to view each as a separate phase of analysis. Once the important and relevant costs and consequences have been identified, they must be measured in appropriate physical and natural units. For example, measurement of the operating costs of a particular screening programme may yield a partial list of *ingredients* such as 500 physical examinations performed by physicians, 10 weeks of salaried nursing time, 10 weeks of a 1000 square foot clinic, 20 hours of medical research librarian time from an adjoining hospital, etc. Similarly, costs borne by patients may be measured, for instance, by the amount of medication purchased, or the number of times travel was required for treatment, or the time lost from work while being treated.

Notice that situations in which resources are jointly used by one or more programmes present a particular challenge to accurate measurement. How much resource use should be allocated to each programme? And on what basis? A common example of this is found in every hospital, where numerous clinical services and programmes share common overhead services (e.g., electric power, cleaning, and administration), provided centrally. In general, there is no nonarbitrary solution to the measurement problem; however, users of results should satisfy themselves that *reasonable* criteria (number of square feet, number of employees, number of cases, etc.) have been used to distribute the common costs. Users should definitely ascertain that such shared costs have been allocated, in fact, to participating services or programmes, as this is a common omission in evaluations! Clinical service directors often argue that small changes in the size of their programmes (up or down) do not affect the consumption of central services. Sometimes it is even argued that overhead costs are unaffected by the service itself. However, though this argument may be intuitively appealing from the viewpoint of a particular programme or service director, the extension of this method to each service in the hospital would imply that the totality of services could be operated without light, heat, power, and secretaries! (The allocation of overhead costs is discussed further in Chapter 4.)

With respect to the measurement of consequences, if the identification of outcomes of interest has been clearly performed, then selection of appropriate units of measurement for programme effects should be relatively straightforward. For example, effects might relate to mortality and be measured in life-years gained or deaths averted; they might relate to morbidity and be measured, for example, in reductions in disability days or improvements on some index of health status measuring physical, social, or emotional functioning; they may be even more specific, depending upon the alternatives under consideration. Thus, percentage increase in weight-bearing ability may be an appropriate natural measurement unit for an evaluation of a physiotherapy programme, while the number of correctly diagnosed cases may be appropriate for a comparison of venography with leg scanning in the diagnosis of deep-vein thrombosis.

Changes in resource use resulting from the effects will be measured in physical units similar to those employed for costs. Thus the changes in utilization resulting from any particular programme will likely be recorded in numbers of procedures, or amounts of time, space, or equipment. Changes in the resource use by patients will continue to be measured in amounts of medication purchased, trips taken for treatment, and so forth.

While the nature of changes in the quality of life may be described, this is one case where measurement in objective, physical or natural units is difficult, although the consequence of some surgical interventions may be quantified in number of complications. However, the adjustment of effects for quality of life is usually a matter of valuation.

6. Were costs and consequences valued credibly?

The sources and methods of valuation of costs, benefits, and utilities should be clearly stated in an economic evaluation. Costs are normally valued in units of local currency, based on prevailing *prices* of, for example, personnel, commodities, and services, and can often be taken directly from programme budgets. All current and future programme costs are normally valued in constant dollars of some base year (usually the present), in order to remove the effects of inflation from the analysis.

It should be remembered that the objective in valuing costs is to obtain an estimate of the worth of resources depleted by the programme. This may necessitate adjustments to some apparent programme costs (e.g., the case of subsidized services or volunteer labour recieved by one programme instead of another). In addition, valuation of the cost of a day of institutional care for a specific condition is particularly troublesome in that the use of an average cost per day (the widely quoted *per diem*), calculated on the basis of the institution's entire annual case load, is almost certainly an overestimate or underestimate of the actual cost for any specific condition, sometimes by quite a large amount. Readers should approach *per diem* values with extreme caution.*

Valuation of *direct benefits* proceeds in the same fashion as for costs and is subject to the same caveats, since those benefits are usually expected future costs that are saved. Valuation of *production gains* or *indirect benefits* (i.e., changes in the value of output of individuals or groups who receive the health programme or service) normally employs individual- or group-wage rates to value the increased working time available. It is here that critics of cost–benefit analyses point out the inequity associated with linking estimates of the value of health programmes to the vagaries of the market. They argue that acceptance of existing wage rates coupled with the inclusion of *indirect benefits* biases cost–benefit studies against programmes aimed at groups such as housewives, the elderly, children, and the unemployed. Although it may be possible to adjust some of the estimates to acknowledge this problem

---

* In principle and (with great effort) in practice, it is possible to identify, measure and value each depleted resource (e.g., drugs, nursing time, light, food, etc.) in treating a specific patient or group of patients. While this yields a relatively accurate cost estimate, the detailed monitoring and data collection are usually prohibitively expensive. The other broad alternative costing strategy is to start with the institution's total costs for a particular period and then to improve upon the method of simply dividing by the total patient-days to produce an average cost-per-day. Quite sophisticated methods of cost allocation to individual hospital departments or wards have been developed, as illustrated by Boyle, Torrance, Horwood, and Sinclair (1982) with respect to neonatal intensive care. An intermediate method involves acceptance of the components of the general *per diem* relating to *hotel* costs (since these are relatively invariant across patients) combined with more precise calculation of the medical treatment costs associated with the specific patients in question. For an example of this intermediate approach see Hull, Hirsh, Sackett, and Stoddart (1982). Of course, the effort devoted to accurate *per diem* estimates depends upon their overall importance in the study; however, unthinking use of *per diems* or average costs should be guarded against.

27

(e.g., imputing a value to housewives' services based on wages for similar work), the *indirect benefit* issue remains controversial.

In valuation of preferences or utilities, we are basically attempting to ascertain how much better the quality of life is in one health situation or 'state' compared with another (e.g., dialysis at home with help from a spouse or friend versus dialysis in hospital). Several techniques are available for making the comparison; the important thing to note is that each will produce an adjustment factor with which to increase or decrease the value of time spent in health situations or 'states', resulting from the alternative in question relative to some baseline. The results of utility analyses are expressed in *healthy days* or *quality-adjusted life-years* gained, as a result of the programmes being evaluated.

Two broad approaches to utility analysis can be found in the literature. The first approach, outlined by Torrance (1982), emphasizes the development of measurement methods and empirical testing on different populations. The other approach, outlined by Weinstein (1981), places emphasis on the estimation of utility values by a quick (and inexpensive) consensus-forming exercise, and then undertaking extensive sensitivity analysis on the chosen values to see whether study results change if the chosen utility estimates are varied. We see a role for both approaches. The latter approach is useful in getting decision-makers to think about resource allocation problems and is, in fact, relatively quick and inexpensive. The measurement approach is useful in highlighting the fact that different actors (doctors, policy-makers, patients and the general public as taxpayers) may have different values, and is clearly crucial in situations where the study result is sensitive to the utility values assigned. (An example of such a case arose in the study by Stason and Weinstein (1977) on the economics of hypertension therapy. The study result was sensitive to whether it was assumed that the side effects of antihypertensive drugs constituted a 1 or 2 per cent reduction in health status.)

Since the measurement of preferences in health is a relatively new field, there are many unresolved issues which readers of cost–utility analyses should note. Users of such analyses will probably want to know, at minimum, whose preferences were used to construct the adjustment factor—the patient's, the provider's, the taxpayer's or the bureaucrat's? If patients' preferences have not been employed, we may want to assure ourselves further that the persons whose preferences did count clearly understood the characteristics of the health state, either through personal experience or through a description of the state presented to them. Many of these issues will be taken up in Chapter 6.

7. Were costs and consequences adjusted for differential timing?

Since comparison of programmes or services must be made at one point in time (usually the present), the timing of programme costs and consequences which do not occur entirely in the present must be taken into account. Different programmes may have different time profiles of costs or consequences. For example, the primary benefits of an influenza immunization programme are immediate while those of hypertension screening occur well into the future. The time profile of costs and consequences may also differ within a single programme; the costs of the hypertension screening programme would be incurred in the present. Therefore, future dollar cost and benefit streams are reduced or 'discounted' to reflect the fact that dollars spent or saved in the future should not weigh as heavily in programme decisions as dollars spent or saved today. This is primarily due to the existence of *time preference*. That is, individually and as a society we prefer to have dollars or resources now as opposed to later because we can benefit from them in the interim. This is evidenced by the existence of interest rates (as well as the popular wisdom about 'a bird in the hand'). Moreover since *time preference* is not exclusively a financial concept, discounting of outcomes should also be undertaken in cost-effectiveness and cost–utility studies (Weinstein and Stason, 1977). The mechanics of discounting and the choice of discount rate are discussed in Chapter 4 (see p. 48).

8. Was an incremental analysis of costs and consequences of alternatives performed?

For meaningful comparison, it is necessary to examine the additional costs that one service or programme imposes over another, compared with the additional effects, benefits, or utilities it delivers. This *incremental* approach to analysis of costs and consequences can be illustrated by reference to one of the examples cited in Chapter 2, namely, strategies for the diagnosis of deep-vein thrombosis (DVT) (Hull *et al*. 1981).

Table 3.1 shows the costs and outcomes (in terms of correct diagnoses*) generated by two alternative strategies: impedance plethysmography (IPG) alone versus IPG plus out-patient venography if impedance plethysmography is negative. (IPG is a noninvasive strategy, whereas venography, the diagnostic 'gold standard' for DVT, can cause

---

* The Hull *et al*. (1981) study is an example of a cost-effectiveness analysis in which the outcomes are not the therapeutic effects themselves, but rather intermediate, diagnostic outcomes with direct implications for therapeutic effects, in that failure to diagnose leads directly to increased morbidity and mortality.

Assessment of economic evaluation

Table 3.1. Economic evaluation of alternative diagnostic strategies for 516 patients with clinically suspected deep-vein thrombosis[a]

Programme	Costs ($ US)	Outcomes (No. of correct diagnoses)	Ratio of cost to outcome ($ per correct diagnosis)
1. IPG (alone)	321 488	142	2264
2. IPG plus out-patient venography if IPG negative	603 552	201	3003
3. Increment (of Programme 2 over Programme 1)	282 064	59	4781

[a] Data drawn from Table 1, Hull *et al.* (1981), by permission.

pain and other unpleasant side effects.) Although one could compare the simple ratios of costs to outcomes for the two alternatives, the correct comparison is the one of incremental costs over incremental outcomes, since this tells us how much we are paying (for each extra correct diagnosis) in adding the extra diagnostic test. In this case the relevant figure is therefore $4781 per correct diagnosis, not the average figure for the second programme, $3003 per correct diagnosis. It may be decided that $4781 is still a price worth paying; however, it is important to be clear on the principle since earlier (in Chapter 2, see p. 7) we pointed out that, in the case of screening for cancer of the colon, there was a big difference between the average cost (per case detected) of a protocol of six sequential tests and the incremental cost of performing a sixth test, having aleady done five (Neuhauser and Lewicki 1975).

Obviously similar analyses could be performed if the consequences were effects in natural units (e.g., years of life) or in utilities (e.g., quality-adjusted life-years).

9. Was a sensitivity analysis performed?

Every evaluation will contain some degree of uncertainty, imprecision or methodological controversy. What if the compliance rate for influenza vaccination was 10 per cent higher than considered for the analysis? What if the *per diem* hospital cost still understated the true resource cost of a treatment programme by $100? What if a discount rate of 6 per cent had been used instead of 2 per cent? Or what if indirect costs and benefits had been excluded from the analysis? Users of efficiency studies will often ask these and similar questions; therefore, careful analysts will identify critical methodological assumptions or areas of uncertainty. Furthermore, they will often attempt to rework the analysis (qualitatively if not quantitatively), employing different assumptions or estimates in order to test the sensitivity of the results and conclusions to such changes. If large variations in the assumptions or estimates underlying an analysis do not produce significant alterations in the results then one would tend to have more confidence in the original results. If the converse occurs, more effort is then required to reduce the uncertainty and/or improve the accuracy of the critical variables. In either case, this *sensitivity analysis* is an important element of a sound economic evaluation. Sensitivity analysis is discussed further in Chapter 5 (see p. 82).

10. Did the presentation and discussion of study results include all issues of concern to users?

It will be clear from the foregoing discussion that the economic analyst has to make many methodological judgements when undertaking a study. Faced with users who may be mainly interested in the 'bottom line'—e.g., 'Should we buy a C–T scanner'—how should he present his results?

Decision indices such as cost-effectiveness and cost–benefit ratios are a useful way of summarizing study results. However, they should be used with care for, in interpreting them, the user may not be completely clear on what has gone into their construction. Some analysts give a range of results. For example, in their economic evaluation of neonatal intensive care for very-low-birth-weight infants, Boyle *et al.* (1983) compare the results for infants below 1000g and from 1000–1500g in terms of costs to hospital discharge, costs and consequences to age 15 and costs and consequences for lifetime (Table 3.2). They leave it to the users of the study to decide on which index (or indices) neonatal intensive care should be judged, since the different measures incorporate different value judgements and varying amounts of precision. (For example, the index of 'net economic benefit' includes production gains/losses, and the index of 'cost per quality-adjusted life-year' incorporates the preferences for health states of a sample of the local population.) This leads to another general point, namely, it is important for the analyst to be as explicit as possible about the various judgements he has made in carrying out the study. A good study should leave the user more (rather than less!) aware of the various technical and value judgements necessary to arrive at resource allocation decisions in health care.

Table 3.2. Measures of economic evaluation of neonatal intensive care according to birth-weight class (5 per cent discount rate)[a]

Period	Birth-weight class	
	1000–1499 g	500–999 g
	$	
To hospital discharge[b]		
Cost/additional survivor at hospital discharge	59 500	102 500
To age 15 (projected)		
Cost/life-year gained	6 100	12 200
Cost/QALY[c] gained	7 700	40 100
To death (projected)		
Cost/life-year gained	2 900	9 300
Cost/QALY[c] gained	3 200	22 400
Net economic benefit (loss)/live birth	(2 600)	(16 100)
Net economic cost/life-year gained	900	7 300
Net economic cost/QALY[c] gained	1 000	17 500

[a] Values are expressed in 1978 Canadian dollars. Multiply by 0.877 to calculate equivalent 1978 U.S. dollars. [b] All costs and effects occurred in year one. [c] QALY denotes quality-adjusted life-year. (From Boyle *et al*. (1983), by permission.)

Finally, a good study should begin to help the user interpret the results in the context of his own particular situation. This can be done by being explicit about the viewpoint for the analysis (an earlier point) and by indicating how particular costs and benefits might vary by location. For example, the costs of instituting day-care surgery may vary, depending on whether a purpose-built day-care unit already exists or whether wards have to be converted. Similarly, the benefits of day-care surgery may vary depending on whether, in a particular location, there is pressure on beds and whether beds will be closed or left empty (Russell, Devlin, Fell, Glass, and Newell 1977). Obviously it is impossible for the analyst to anticipate every possibility in every location, but one limitation of economic evaluation techniques (discussed in Section 3.2) is that they assume that freed resources will be put to other beneficial uses. Evans and Robinson (1980) argue that in the case of day-care surgery the full economic payoff may not have been obtained in at least one Canadian hospital.

3.2 LIMITATIONS OF ECONOMIC EVALUATION TECHNIQUES

Our main purpose in this chapter is to make the user of economic evaluation results more aware of the methodological judgements involved in undertaking an economic evaluation in the health care field. In Annex 3.1 we have consolidated the points made in this chapter into a suggested checklist of questions to ask when critically assessing economic evaluation results. Some of these questions signal limitations of economic evaluation techniques. For example, economic evaluation techniques assume, rather than establish, programme effectiveness. In addition, there are several other limitations of which users should be aware.

Of primary concern from a policy viewpoint is the fact that economic evaluations do not usually incorporate the importance of the distribution of costs and consequences into the analysis. Yet, in some cases, the identity of the recipient group (e.g., the poor, the elderly, working mothers, or a remote community) may be an important factor in assessing the social desirability of a service or programme. Indeed, it may be the motivation for the programme in the first place. Although it is sometimes suggested that differential weights be attached to the value of outcomes accruing to special recipient groups, this is not normally done within an economic evaluation. Rather, an equitable distribution of costs and consequences across socioeconomic or other defined groups in society is viewed as a competing dimension upon which decisions are made, in addition to that of efficient deployment of resources.

It should also be noted that economic evaluation techniques assume that resources freed or saved by preferred programmes will not in fact be wasted but will be employed in alternative worthwhile programmes. This assumption warrants careful scrutiny, for if the freed resources are consumed by other, ineffective or unevaluated, programmes, then not only is there no saving, but overall health system costs will actually increase without any assurance of additional improvements in the health status of the population.

Finally, evaluation of any sort is in itself a costly activity. Bearing in mind that *even a cost–benefit analysis should be subject to a cost–benefit analysis*, it seems reasonable to suggest that economic evaluation techniques will prove most useful in situations where programme objectives require clarification, the competing alternatives are significantly different in nature, or large resource commitments are under consideration.

3.3. CONCLUSIONS

In these introductory chapters, we have tried to assist users of economic evaluations in interpreting evaluation studies and assessing their usefulness for health care decisions, or for planning further analyses. The rationale for economic evaluation, its fundamental characteristics and the basic types of economic evaluation were described in Chapter 2. In Chapter 3 we have identified and discussed 10 questions which readers of economic evaluations can ask in order to critically assess a particular study; a checklist of these questions is given in Annex 3.1.

Our intent in offering a checklist is not to create hypercritical users who will be satisfied only by superlative studies. It is important to realize, as emphasized at the outset, that for a variety of reasons it is unlikely that every study will satisfy all criteria. However, the use of these criteria as screening devices should help users of economic evaluations to identify quickly the strengths and weaknesses of studies. Moreover, in assessing any particular study, users should ask themselves one final question, *'How does this evaluation compare with our normal basis for decision-making?'* They may find that the method of organizing thoughts embodied in the evaluation compares well with alternative approaches, even bearing in mind the possible deficiencies in the study.

REFERENCES

Boyle, M. H., Torrance, G. W., Horwood, S. P., and Sinclair, J. C. (1982). *A cost analysis of providing neonatal intensive care to 500–1499 gram birth-weight infants*, Research Report No. 51, Programme for Quantitative Studies in Economics and Population. McMaster University, Hamilton, Ontario.

Boyle, M. H., Torrance, G. W., Sinclair, J. C., and Horwood, S. P. (1983). Economic evaluation of neonatal intensive care of very low birth-weight infants. *N. Engl. J. Med.* **308**, 1330–7.

Department of Clinical Epidemiology and Biostatistics (1981). How to read clinical journals. V: To distinguish useful from useless or even harmful therapy. *Can. Med., Assoc. J.* **124**, 1156–62.

Drummond, M. F. (1981). Welfare economics and cost–benefit analysis in health care. *Scottish Journal of Political Economy* **28**, 125–45.

Evans, R. G. and Robinson, G. C. (1980). Surgical day care: measurements of the economic payoff. *Can. Med. Assoc. J.* **123**, 873–80.

Hull, R., Hirsh, J., Sackett, D. L., and Stoddart, G. L. (1981). Cost-effectiveness of clinical diagnosis, venography and noninvasive testing in patients with symptomatic deep-vein thrombosis. *N. Engl. J. Med.* **304**, 1561–7.

——, ——, ——, ——. (1982). Cost-effectiveness of primary and secondary prevention of pulmonary embolism in high-risk surgical paients. *Can. Med. Assoc. J.* **127**, 990–5.

Neuhauser, D. and Lewicki, A. M. (1975). What do we gain from the sixth stool guaiac? *N. Engl. J. Med.* **293**, (5), 226–8.

Russell, I. T.., Devlin, H. B., Fell, M., Glass, N. J., and Newell, D. T. (1977). Day case surgery for hernias and haemorrhoids: a clinical, social and economic evaluation. *Lancet* **i**, 844–7.

Stason, W. B. and Weinstein, M. D. (1977). Allocation of resources to manage hypertension. *N. Engl. J. Med.* **296**, 732–9.

Stoddart, G. L. (1982). Economic evaluation methods and health policy. *Evaluation and the Health Professions* **5**, (4), 393–414.

Torrance, G. W. (1982). Preferences for health states: A review of measurement methods. In *Clinical and Economic Evaluation of Perinatal Programmes* (ed. J. C. Sinclair). Proceedings of Mead Johnson Symposium on Perinatal and Developmental Medicine, **20**, 37–45, Vail, Colorado, June 6–10.

Weinstein, M. C. (1981). Economic assessments of medical practices and technologies. *Medical Decision Making* **1** (4), 309–30.

—— and Stason, W. B. (1977). Foundations of cost-effectiveness analysis for health and medical practices. *N. Engl. J. Med.* **296**, 716–21.

Weisbrod, B. A., Test, M. A., and Stein, L. I. (1980). Alternative to mental hospital treatment. II. Economic benefit–cost analysis. *Arch. General Psychiatry* **37**, 400–5.

Williams, A. H. (1981). Welfare economics and health status measurement. In *Health, Economics and Health Economics* (ed. J. van der Gaag and M. Perlman) pp. 271–81. Amsterdam, North-Holland.

ANNEX 3.1. A SUGGESTED CHECKLIST FOR ASSESSING ECONOMIC EVALUATIONS

1. Was a well-defined question posed in answerable form?

    1.1 Did the study examine both costs and effects of the service(s) or programme(s)?

    1.2 Did the study involve a comparison of alternatives?

    1.3 Was a viewpoint for the analysis stated and was the study placed in any particular decision-making context?

2. **Was a comprehensive description of the competing alternatives given? (i.e., can you tell who? did what? to whom? where? and how often?)**

    2.1 Were any important alternatives omitted?

    2.2 Was (Should) a *do-nothing* alternative (be) considered?

3. **Was there evidence that the programmes' effectiveness had been established?**

    3.1 Has this been done through a randomized, controlled clinical trial? If not, how strong was the evidence of effectiveness?

4. **Were all the important and relevant costs and consequences for each alternative identified?**

    4.1 Was the range wide enough for the research question at hand?

    4.2 Did it cover all relevant viewpoints? (Possible viewpoints include the community or social viewpoint, and those of patients and third party payers. Other viewpoints may also be relevant depending upon the particular analysis.)

    4.3 Were capital costs, as well as operating costs, included?

5. **Were costs and consequences measured accurately in appropriate physical units? (e.g., hours of nursing time, number of physician visits, lost workdays, gained life-years)**

    5.1 Were any of the identified items omitted from measurement? If so, does this mean that they carried no weight in the subsequent analysis?

    5.2 Were there any special circumstances (e.g., joint use of resources) that made measurement difficult? Were these circumstances handled appropriately?

6. **Were costs and consequences valued credibly?**

    6.1 Were the sources of all values clearly identified? (Possible sources include market values, patient or client preferences and views, policy-makers' views and health professionals' judgements.)

    6.2 Were market values employed for changes involving resources gained or depleted?

    6.3 Where market values were absent (e.g., volunteer labour), or market values did not reflect actual values, (such as clinic space donated at a reduced rate), were adjustments made to approximate market values?

6.4 Was the valuation of consequences appropriate for the question posed? (i.e., Has the appropriate type or types of analysis—cost-effectiveness, cost–benefit, cost–utility—been selected?)

7. Were costs and consequences adjusted for differential timing?

7.1 Were costs and consequences which occur in the future 'discounted' to their present values?

7.2 Was any justification given for the discount rate used?

8. Was an incremental analysis of costs and consequences of alternatives performed?

8.1 Were the additional (incremental) costs generated by one alternative over another compared to the additional effects, benefits or utilities generated?

9. Was a sensitivity analysis performed?

9.1 Was justification provided for the ranges of values (for key study parameters) employed in the sensitivity analysis?

9.2 Were study results sensitive to changes in the values (within the assumed range)?

10. Did the presentation and discussion of study results include all issues of concern to users?

10.1 Were the conclusions of the analysis based on some overall index or ratio of costs to consequences (e.g., cost-effectiveness ratio)? If so, was the index interpreted intelligently or in a mechanistic fashion?

10.2 Were the results compared with those of others who have investigated the same question?

10.3 Did the study discuss the generalizability of the results to other settings and patient/client groups?

10.4 Did the study allude to, or take account of, other important factors in the choice or decision under consideration (e.g., distribution of costs and consequences, or relevant ethical issues)?

Assessment of economic evaluation

10.5 Did the study discuss issues of implementation, such as the feasibility of adopting the 'preferred' programme given existing financial or other constraints, and whether any freed resources could be redeployed to other worthwhile programmes?

4. Cost analysis

4.1. SOME BASICS

The analysis of the comparative costs of alternative treatments or health care programmes is common to all forms of economic evaluation and therefore most of the methodological issues discussed in this chapter are likely to be of relevance to all analyses. Two particularly thorny issues, the treatment of overhead costs (techniques for allocating shared overhead costs to individual projects) and allowance for differential timing of costs (the techniques of discounting and annuitization of capital expenditure), will be discussed in some detail. However, the chapter begins by covering some of the basic questions that an evaluator might have when embarking on a costing study in the health field.

4.1.1. Which costs should be considered?

The main categories of costs of health care programmes or treatments were identified in Fig. 3.1 of Chapter 3; these are the organizing and operating costs within the health sector, costs borne by patients and their families, and costs borne externally to the health sector, patients, and their families. The particular range of costs included in a given study is likely to be decided upon as a result of considering the following four points.

1. *What is the viewpoint for the analysis?*

It is essential to specify the viewpoint since an item may be a cost from one point of view, but not a cost from another. (For example, patients' travel costs are a cost from the patients' point of view and from the point of view of society, but not a cost from the Ministry of Health's point of view. Workmen's compensation payments are a cost to the paying government, a gain to the patient (recipient), and neither a cost nor a gain to society. (These money transfers, which do not reflect resource consumption, are called transfer payments by economists. Costs are involved in their administration, but these are not measured by the amounts themselves.)

Cost analysis

Possible points of view include: society, Ministry of Health, other provincial ministries, total provincial government, patient, employer, federal government, the agency providing the programme, etc. If the evaluation is being commissioned by a given body, this may give a clue to the relevant point(s) of view. However, when in doubt always adopt the societal point of view, which is the broadest one and is always relevant.

2. *Is the comparison restricted to the two or more programmes immediately under study?*

If the comparison is restricted to the programmes or treatments immediately under study, costs common to both need not be considered as they will not affect the choice between the given programmes. (Elimination of such costs can save the evaluator a considerable amount of work.) However, if it is thought that at some later stage a broader comparison may be contemplated, including other alternatives not yet specified, it might be prudent to consider all the costs of the programmes.

3. *Are some costs merely likely to confirm a result that would be obtained by consideration of a narrower range of costs?*

Sometimes the consideration of patients' costs merely confirms a result that might be obtained from, say, consideration of only operating costs within the health sector. Therefore, if consideration of patients' costs requires extra effort and the choice of programme would not be changed, it may not be worthwhile to complicate the analysis unnecessarily. However, some justification for such an exclusion of a cost category should be given.

4. *What is the relative order of magnitude of costs?*

It is not worth investing a great deal of time and effort considering costs that, because they are small, are unlikely to make any difference to the study result. However, some justification should be given for the elimination of such costs, perhaps based on previous empirical work. It is still worthwhile identifying such cost categories in any event, although the estimation of them might not be pursued in any great detail.

Above all, the main point to remember when embarking on a costing study is that, to an economist, cost refers to the sacrifice (of benefits) made when a given resource is consumed in a programme or treatment. Therefore, it is important not to confine one's attention to expenditures, but to consider also other resources, the consumption of which is not adequately reflected in market prices, e.g., volunteer time, patients' leisure time, donated clinic space, etc.

4.1.2. How should costs be estimated?

Once the relevant range of costs has been identified the individual items must be measured and valued. In general, the programme ingredients approach suggested in Chapter 3 should suffice and market prices will be readily available for many of the cost items. Although the theoretical proper price for a resource is its opportunity cost (i.e., the value of the forgone benefits because the resource is not available for its best alternative use), the pragmatic approach to costing is to take existing market prices unless there is some particular reason to do otherwise (e.g., the price of some resources may be subsidized by a third party such as a charitable institution).

Although the costing of most resource items is relatively unambiguous, the following five issues commonly arise in costing studies.

1. *How are values imputed for nonmarket items?*

The major nonmarket resource inputs to health care programmes are volunteer time and patient/family leisure time. One approach to the valuation of these would be to use market wage rates (e.g., for volunteer time one might use unskilled wage rates). The market value of leisure time is harder to assess. One can argue for a value of lost leisure time of anything from zero, through average earnings, to average overtime earnings (time and a half or double time). The argument for the overtime rate is that this is the price that an employer must pay, at the margin, to buy some of the worker's leisure time. The most common practice is to value lost leisure time at zero in the base case analysis, and to investigate the impact of the other assumptions through sensitivity analysis.

A slightly different approach is to identify and measure units of, say, volunteer input and to document these alongside the other costs when reporting results. This would enable the decision-maker to note those programmes relying heavily on volunteers. It would then be up to the programme director to demonstrate that such an input could be obtained without an opportunity cost to other programmes arising from the diversion of volunteers to the new programme.

The valuation of nonmarket items is discussed further in Chapter 7 on cost–benefit analysis.

2. *How should capital outlays (on equipment, buildings and land) be handled?*

Capital costs are the costs to purchase the major capital assets required by the programme: generally equipment, buildings and land. Capital costs differ from operating costs in a number of ways. First, they represent investments at a single point in time, often at the beginning of the programme, rather than annual sums like operating costs. Frequently, the capital costs are not listed in the accounts or budgets of the organization because they have been funded in advance, perhaps by a one-time grant, while the budgets and accounts represent operating expenses only. Sometimes, the annual budgets and accounts contain an item called depreciation which relates to capital costs, as explained below.

Capital costs represent an investment in an asset which is used over time. Most assets, such as equipment and buildings, wear out, or depreciate, with time. On the other hand, land is a non-depreciable asset because it maintains its value. There are two components of capital cost. One is the opportunity cost of the funds tied up in the capital asset. This is clearly seen in the case of land. Although an investment in non-depreciable land will return the original capital sum when sold, there is still a 'cost'. This cost is the lost opportunity to invest the sum in some other venture yielding positive benefits. This is called the opportunity cost and is valued by applying an interest rate (equal to the discount rate used in the study) to the amount of capital invested.

The second component of a capital cost represents the depreciation over time of the asset itself. Various accounting procedures (straight line, declining balance, double declining balance, etc.) are available for use in the accounts of the organization. Often, accounting practices relate more to the company tax laws governing the depreciation of assets than to the real change in the value of the asset.

There are several methods of measuring and valuing capital costs in an economic evaluation. The best method is to annuitize the initial capital outlay over the useful life of the asset; that is, to calculate the 'equivalent annual cost'. This method and its advantages are discussed in more detail by Richardson and Gafni (1983). The method automatically incorporates both the depreciation aspect and the opportunity cost aspect of the capital cost. It is our preferred approach and is described in Section 4.2 below. An alternative but less exact method is to determine the depreciation cost each year using an accounting method and to determine the opportunity cost on the undepreciated balance for each year (See Levin 1975, Boyle, Torrance, Horwood, and Sinclair 1982). Where market rates exist for the rental of buildings or lease of equipment, these may be used to estimate capital costs. This method also incorporates both the depreciation and the opportunity components of the cost. (A series of exercises illustrating the different methods of measuring and valuing capital costs is given in Annex 4.1.)

If capital outlays relate to resources that are used by more than one programme they may require allocation in a similar fashion to 'overhead' costs. See the discussion of this point below.

3. *What is the significance of the average cost/marginal cost distinction?*

Economists tend to emphasize this point, and the example of the sixth stool guaiac in Chapter 2 illustrated the pitfalls in making decisions based on average cost. In fact, marginal cost and average cost are but two concepts relating costs to quantity (Horngren 1982). A longer list would comprise:

Total cost (TC) = cost of producing a particular quantity of output.

Fixed cost (FC) = costs which do not vary with the quantity of output in the short run (about one year), e.g. rent, equipment lease payments, some wages and salaries. That is, costs which vary with time, rather than quantity.

Variable cost (VC) = costs which vary with the level of output, e.g. supplies, food, fee for service.

Cost function (TC) $= f(Q)$, total cost as a function of quantity.

Average cost (AC) $=$ TC$/Q$, the average cost per unit of output.

Marginal cost (MC)$=$ (TC of $x + 1$ units) $-$ (TC of x units).

       = d(TC)/dQ evaluated at x

       = the *extra* cost of producing *one* extra unit of output.

The major significance of the averge-cost/marginal-cost distinction to the evaluator is as follows. First, when making a comparison of two or more programmes, it is worth asking independently of each, 'What would be the costs (and consequences) of having a little more or a little less?' [e.g., suppose Neuhauser and Lewicki (1975) had been comparing the six-stool protocol for detecting colonic cancer with another diagnostic test. Perhaps the question of six- versus five-tests may never have been asked!] Second, when examining the effects (on cost) of small changes in output, it is likely that these will differ from average costs. For example, the extra cost of keeping a patient in hospital for another day at the end of his treatment might be less than the average daily cost for his whole stay. (In fact, this issue usually arises in the opposite sense—the savings from a reduction of one day's stay are usually lower than the average daily cost.)

4. *How should shared (or overhead) costs be handled?*

The term *overhead costs* is an accounting term for those resources that serve many different departments and programmes, e.g. general hospital administration, central laundry, medical records, cleaning, porters, power, etc. If individual programmes are to be costed, these shared costs may need to be attributed to programmes.

The main point to note at the outset is that there is no unambiguously *right* way to apportion such costs. The approach that is favoured by economists is to employ marginal analysis. That is, to see which (if any) of such costs would change if a given programme were added to, or subtracted from, the overall activity. Whilst this is fine up to a point, the most common situation is that the choice is not such an addition or subtraction, but one between two programmes, each of which would consume the given central services (perhaps because they are competitors for the same space in the hospital). For example, suppose the question concerned space in the hospital that could be used either for anticoagulant therapy for pulmonary embolism, or for renal dialysis. If the economic evaluation concerned a choice between these two programmes, then there would be no methodological problem, the costs associated with use of the space would be common to both and could be excluded from the analysis. However, typically the comparison might be between the anticoagulant therapy and another programme in the same field. This could be a programme of more definitive diagnosis of pulmonary embolism, which would avert some hospitalization. In such an instance it would be relevant to obtain an estimate of the value of the freed resources (e.g. hospital floor space) that could be diverted to other uses.

Essentially, the issue at stake here is that of accurately estimating all the costs attributable to a given programme or treatment when this is delivered alongside other programmes, as in the acute hospital. In Chapter 3 the reader was warned against the unthinking use of hospital (or other institutional) *per diems* or average costs. Before the methods available for apportioning institutional costs are described in more detail, the dangers of using *per diems* require more elucidation.

Many institutions calculate a *per diem* or average cost of their operations. This is essentially their total operating costs for the year divided by their total patient utilization for the year. A common example is a hospital's average cost per patient-day. It is tempting simply to multiply this figure by the number of patients and their average length of stay to determine the hospital cost of a programme. What is wrong with this procedure? First, it is only valid for truly 'average' patients—that is, patients who use an average amount of radiological services, laboratory services, operations, nursing attention, drugs, and so on. If patients in the programme being costed are not average, the result will be in error.

Second, many *per diem* calculations include arbitrary adjustments. For example, certain types of patients (outpatients, day patients, newborn patients, etc.) may be excluded from the denominator of the calculation in recognition that they are not typical. Then an estimate of the costs of these patients (often a very crude estimate) is subtracted from the numerator before calculating the *per diem*. The result is that the *per diem* itself is imprecise, even for the truly average patient.

Finally, typical *per diem* cost figures are incomplete, as they totally ignore capital costs. In summary, *per diem* costs are only applicable to average patients and even then are imprecise and incomplete.

A number of methods can be used to determine a more accurate cost of a programme in a hospital or other setting where shared (or overhead) costs are involved. The methods are illustrated below in terms of a hospital setting. The basic idea is to determine the quantities of service consumed by the patient (days of stay in ward A, B, or C, number of laboratory tests of each type, number of radiological procedures, number of operations, etc.), to determine a full cost (including the proper share of overhead, capital, etc.) for a unit of each type of service, and to multiply these together and sum up the results. The allocation methods described below are different ways to determine the cost per unit for each type of service. In these methods the overhead costs (e.g., housekeeping) are allocated to other departments (e.g., radiology) on the basis of some measure, called an *allocation basis*, judged to be related to usage of the overhead item (e.g., square feet of floor space in the radiology department might be used to allocate housekeeping costs to radiology).

In deciding which of the following approaches to use, the comments made in Section 4.1.1 above, should be borne in mind. That is, the more important the cost item is for the analysis, the greater the effort that should be made to estimate it accurately. There may conceivably be evaluations for which simple *per diem* costs will suffice, since the result is unlikely to change irrespective of the figure assumed for the cost of hospital care. However, we suspect that such situations are in the minority, given the relative order of magnitude of hospital costs compared to other elements of health care expenditures.

Alternatively, the intermediate approach suggested by Hull, Hirsh, Sackett, and Stoddart (1982) may suffice. Here the *per diem* cost is purged of any items relating to medical care costs, leaving just the 'hotel' component of hospital expenditure. It is then assumed that all patients are 'average' in respect of their hotel costs and that this expenditure can therefore be apportioned on the basis of patient days. Thus, the hotel cost can be calculated for the patients in the programme of interest and combined with the medical care costs attributable to those patients to give the total costs of the programme. (The medical care costs would be estimated separately, using data specifically relating to the patients in the programme.)

Cost analysis

If a more detailed consideration of costs is required, various methods for allocating shared (or overhead) costs are available, namely:

(a) *Direct allocation* (ignores interaction of overhead departments). Each overhead cost (e.g., central administration, housekeeping) is allocated directly to final cost centres (e.g., programmes like day surgery; departments like wards or radiology). Programme X's allocated share of central administration is equal to central administration cost times Programme X's proportion of the allocation basis (say, paid hours). Note, Programme X's proportion is Programme X's paid hours divided by total paid hours of all final cost centres, not total paid hours for the whole organization. The latter method would underestimate the costs in all final cost centres.

(b) *Step down allocation* (partial adjustments for interaction of overhead departments). The overhead departments are allocated in a stepwise fashion to all of the remaining overhead departments and to the final cost centres.

(c) *Step down with iterations* (full adjustment for interaction of overhead departments). The overhead departments are allocated in a stepwise fashion to all of the other overhead departments and to the final cost centres. The procedure is repeated a number of times (about three) to eliminate residual unallocated amounts.

(d) *Simultaneous allocation* (full adjustment for interaction of overhead departments). This method uses the same data as (b) or (c) but it solves a set of simultaneous linear equations to give the allocations. It gives the same answer as method (c) but involves less work. (The method is shown diagrammatically in Fig. 4.1.)

An example showing the different approaches to the allocation of overhead costs is presented in Section 4.3. Further details are available in Horngren (1982), Clements (1974), Kaplan (1973), and Boyle, *et al* (1982).

The effort that one would put into overhead cost allocation would depend on the likely importance of overhead costs (in quantitative terms) for the whole analysis. A much simpler, but cruder, approach is to

(a) identify those hospital costs unambiguously attributable to the treatment or programme in question (e.g., physicians' fees, laboratory tests, drugs). (These are known as the directly allocatable costs.) Allocate these directly and immediately to the programme, then;

(b) deduct, from total hospital operating expenses, the cost of depart-

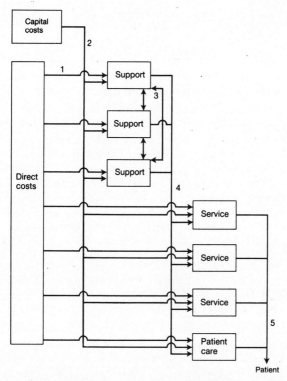

(1) Direct costs assigned directly to cost centres. (2) Capital costs assigned to cost centres. (3) Support cost centres simultaneously allocated to each other. (4) Support cost centres allocated to other cost centres. (5) Costs assigned to each patient based on services used.

Fig. 4.1. Schematic illustration of cost allocations (from Boyle *et al.* 1982)

ments already allocated above and departments known not to service the programme being costed, then;

(c) allocate the remainder of hospital operating expenses on the basis of number of patient days, e.g.:

$$
\begin{array}{l}
\text{Hospital cost} \\
\text{of the} \\
\text{programme}
\end{array}
=
\begin{array}{l}
\text{Directly} \\
\text{allocatable} \\
\text{costs}
\end{array}
+
\frac{\text{Net hospital expenditure}}{\begin{array}{l}\text{Total number} \\ \text{of hospital} \\ \text{patient-days}\end{array}}
\times
\begin{array}{l}
\text{Hospital} \\
\text{patient-days} \\
\text{attributable} \\
\text{to the} \\
\text{programme}
\end{array}
$$

47

(d) finally, undertake a sensitivity analysis.

Whilst there is nothing to suppose that this method is anything but crude, if the choice between programmes is fairly insensitive to the value derived it may suffice.

5. *How should indirect costs be estimated?*

As was mentioned in Chapter 3, this is a particularly contentious issue. The discussion of this point will be postponed until Chapter 5 since changes in productive output more often enter into the economic evaluation as a consequence of health care programmes, that is, the therapy often averts future production losses in that it enables the sick person to return to work or work until later in life. Production losses occur less often on the cost side of the equation since the patient is already off work because of his or her condition. Exceptions here would include population screening or other preventive programmes and anyone considering an evaluation of these should consult the relevant section in Chapter 5.

4.2. ALLOWANCE FOR DIFFERENTIAL TIMING OF COSTS (DISCOUNTING AND THE ANNUITIZATION OF CAPITAL EXPENDITURES)

As was mentioned in Chapter 3, some allowance needs to be made for the differential timing of costs and consequences. That is, even in a world with zero inflation and no bank interest, it would be an advantage to receive a benefit earlier or to incur a cost later—it gives you more options. Economists call this the notion of *time preference*.

Typically, economic evaluation texts discuss the situation where the costs of the alternative programmes A and B can be identified by the year in which they occur:

Year	*Cost of Programme A* (*$000s*)	*Cost of Programme B* (*$000s*)
1	5	15
2	10	10
3	15	4

In this example, B might be a preventive programme which requires more outlay in Year 1 with the promise of lower cost in Year 3. The crude addition of the two cost streams shows B to be of lower cost, but the outlays under A occur more in the later years.

A comparison of A and B (adjusted for the differential timing of resource outlays) would be made by discounting future costs to present values. The calculation is performed as follows. If P = present value; F_n = future cost at year n; and r = annual interest (discount) rate (e.g., 0.05 or 5 per cent), then

$$P = \sum_{n=1}^{3} F_n(1+r)^{-n}$$

$$= \frac{F_1}{(1+r)} + \frac{F_2}{(1+r)^2} + \frac{F_3}{(1+r)^3}$$

In our example this gives:

Present value of cost of A = 26.79

Present value of cost of B = 26.81

This assumes that the costs all occur at the end of each year. An alternative assumption which is commonly used is to assume that the costs all occur at the beginning of each year. Then, Year 1 costs need not be discounted, Year 2 costs should be discounted by one year, etc. Calculated in this way, the previous example is:

$$P = \sum_{n=0}^{2} F_n(1+r)^{-n}$$

$$= F_0 + \frac{F_1}{(1+r)} + \frac{F_2}{(1+r)^2}$$

Present value of A = 28.13

Present value of B = 28.15

The factor $(1+r)^{-n}$ is known as the discount factor and can be obtained for a given n and r from Table 1 in Annex 4.2. For example, the discount factor for three periods (years) at a discount rate of 5 per cent is 0.8638.

While this approach is the most convenient for a number of programme comparisons, a more common situation is that where most of the costs are easily expressed on an annual recurring basis and it is only capital costs which differ from year to year (typically these will be at the beginning of the programme, or Year 0).

Here it might be more convenient to express all the costs on an annual

Cost analysis

basis, obtaining an equivalent annual cost (E) for the capital outlay by an amortization or annuitization procedure. This works as follows:

If the capital outlay is K, we need to find the annual sum E which over a period of n years (the life of the facility), at an interest rate of r, will be equivalent to K.

This is expressed by the following formula:

$$K = \frac{E}{(1+r)} + \frac{E}{(1+r)^2} + \ldots + \frac{E}{(1+r)^n}$$

$$K = E \frac{1 - (1+r)^{-n}}{r}$$

$K = E$ [Annuity factor, n periods, interest r]

As before, the annuity factor is easily obtainable from Table 2 in Annex 4.2. For example, in the cost analysis of providing long term oxygen therapy Lowson, Drummond, and Bishop (1981) found the total capital (set up) costs (K) to be £2153.

Therefore, applying the formula given above:

$$2153 = \frac{E}{(1+r)} + \frac{E}{(1+r)^2} + \frac{E}{(1+r)^3} + \frac{E}{(1+r)^4} + \frac{E}{(1+r)^5}$$

$2153 = E$ [Annuity factor, 5 years, interest rate 7 per cent]

$2153 = E[4.1002]$ (from Table 2 in Annex 4.2)

$E = £525$ (as shown in Table III of Lowson *et al.*, (1981).

Note that Lowson *et al* (1981) assumed that the annuity was in arrears, that is, due at the end of the year. It might be argued that a more realistic assumption would be that it were payable in advance. This is equivalent to the formula:

$$2153 = \frac{E}{(1+r)} + \frac{E}{(1+r)^2} + \frac{E}{(1+r)^3} + \frac{E}{(1+r)^4}$$

The value for E can still be obtained from Table 2 by taking one less period and adding 1.000. This gives a lower value for $E = £491$. This is logical since the repayments are being made earlier (at the beginning of each year) rather than in arrears.

This approach can be generalized to handle the situation where the equipment or buildings have a resale value at the end of the programme. If

$S =$ the resale value;
$n =$ the useful life of the equipment;
$r =$ discount (interest rate);
$A(n, r) =$ the annuity factor (n years at interest rate r);
$K =$ purchase price/initial outlay;
$E =$ equivalent annual cost;

then

$$E = \frac{K - \dfrac{S}{(1+r)^n}}{A(n,r)}$$

The method described above is unambiguous for new equipment. For old equipment, there are two choices:

    Choice 1 — Use the replacement cost of the equipment (or the original cost indexed to current dollars) and a full life.
    Choice 2 — Use the current market value of the old machine and its remaining useful life.

Choice 1 is usually better as the results are more generalizable—less situational. Note that using the undepreciated balance from the accounts of the organization is never a method of choice.

It can be seen that the equivalent annual cost of buildings or equipment to a given programme depends on the values of n, r, and S, all of which must be assumed at the time of the evaluation. Practical points that evaluators might care to note are:

1. *Useful life and resale value (*n *and* S)

It is important to make a distinction between the physical life of a piece of equipment and its useful clinical life. The latter is highly dependent on technological change. Obviously one can undertake a sensitivity analysis using different values for n, but in general it is best to be conservative and assume short lives (say, around five years) for clinical equipment.

2. *Choice of the discount rate (*r)

There are *two competing theories* regarding the proper measure for the discount rate for public projects (the social discount rate):

    (a) $r =$ the real rate of return (to society) forgone in the private sector.

Cost analysis

This can be estimated empirically, although not without controversy.

(b) r = the social rate of time preference.

The social rate of time preference is a measure of society's willingness, collectively, to forgo consumption (gratification) today in order to have greater consumption (gratification) tomorrow. Frequently it is assumed that the interest rate on a risk free investment (e.g., long-term government bonds) represents the individual investor's willingness to forgo the present for the future, and that this rate is the individual's rate of time preference. Then if society's collective rate of time preference is simply the aggregate of the individual rates (a controversial assumption), the required rate is simply given by the real (adjusted for inflation) rate of return on long-term government bonds.

However, in practice it is usually admissible to select a central 'best estimate' of r, and then vary this systematically in a sensitivity analysis to determine the impact on the study conclusions. The criteria to use in selecting a central r and a range for sensitivity analysis are that these should:

(a) be consistent with economic theory (2 per cent to 10 per cent);
(b) include (bracket) any government recommended rates (5 per cent, 7 per cent, 10 per cent);
(c) include (bracket) rates that have been used in other published studies to which you might wish to compare results (3 per cent to 10 per cent);
(d) be consistent with 'current practice' (for example, 5 per cent has been used recently in papers published in the *New England Journal of Medicine*).

3. *How to handle inflation*

If it is assumed that all the items of cost in the programme will inflate at the same rate and that this will be the same rate as inflation in general, there are two equivalent choices:

(a) Inflate all future costs by this predicted inflation rate and then use a larger discount rate that allows for the effect of general inflation (the inflation adjusted discount rate*), or

---

* Calculation of inflation adjusted discount rate: if the real discount rate is 5 per cent and general inflation is 8 per cent, then the inflation adjusted $r = (1.05)(1.08) = 1.134$ or 13.4 per cent.

(b) Do not inflate any future costs (i.e., use constant dollars) and use a smaller discount rate that does not allow for inflation (the real discount rate).

Method (b) is the simpler and preferred approach.

If it is assumed that different items of cost in the programme will inflate at different rates, there are also two equivalent choices:

(a) Inflate all future costs by their particular predicted inflation rates and then use a larger discount rate that allows for the effect of general inflation (the inflation adjusted discount rate*), or

(b) Do not inflate any future costs (i.e., use constant dollars) and use a smaller discount rate that does not allow for inflation (the real discount rate), but adjust the discount rate for each item to account for the differential inflation rate between this item and the 'general' rate of inflation, e.g., if general inflation is 8 per cent, this item is expected to inflate 10 per cent, and the real r is equal to 4 per cent, then r adjusted for this item is

$$r = 1.04 \times \frac{1.08}{1.10} = 1.021, \text{ i.e. } 2.1 \text{ per cent.}$$

Method (b) is again the preferred approach. In general, however, most studies perform the whole analysis in constant price terms and use a single discount rate.

4.3. ALLOCATION OF OVERHEAD COSTS: EXAMPLE

The following example demonstrates the various methods of handling overhead costs discussed in Section 4.1.2 (4), p. 43. Suppose we wish to determine the cost of neonatal intensive care (NIC) for a specific group of patients. For each patient we have data on the length of stay in the neonatal intensive care unit (NICU) and data on the number and type of laboratory tests performed. For simplicity, let us assume that these were the only services received by the patients—that is, the patients had no operations, no radiological or nuclear medicine investigations, no social work, etc. Furthermore, let us assume that there are only three overhead departments that serve the laboratory and the NICU: administration, housekeeping and laundry. (In principle it would be possible to consider other overhead departments, like plant operations and maintenance, bioengineering, and materials management.)

* See footnote on p. 52.

Cost analysis

The first task is to determine a unit of output for those departments that directly serve patients. We will be determining a cost per unit of output, and multiplying this cost by the usage of each patient to determine the cost per patient. Thus, the unit of output must be as homogeneous as possible with respect to cost, and yet be available in the data for each patient. We have selected a *patient-day* as the unit of output of the NICU, and a *DBS unit* for the laboratory. A DBS (Dominion Bureau of Statistics) unit is a standard laboratory work unit used in Canada; each lab test is assigned a predetermined number of DBS units according to the amount of work needed to perform the test.

An allocation basis must be determined for each overhead department. For example, square feet of floor space has been selected for housekeeping. This means that housekeeping costs will be allocated to departments receiving housekeeping services in proportion to the square footage of floor space in the department. Similarly, paid hours has been selected as the allocation basis for administration costs, and pounds of laundry for the laundry costs.

The data for this simplified example are given in Table 4.1. The calculations, as performed by the different methods, are given in Tables 4.2 to 4.7.

Table 4.1. Cost allocation data

	Annual direct cost[a] $	Annual units of output[b]	Direct cost per unit $	Allocation basis	Annual pd-hrs	Ft^2	Annual lbs laundry
Overhead departments							
Administration	2 000 000			pd-hrs	200 000	30 000	0
Housekeeping	1 500 000			ft^2	300 000	4 000	80 000
Laundry	1 300 000			lbs	200 000	8 000	0
Other	10 200 000				300 000	158 000	120 000
Subtotal	*15 000 000*				*1 000 000*	*200 000*	*200 000*
Final departments (Pt. service)							
Laboratory	4 000 000	8 000 000	0.50/DBS unit		250 000	30 000	25 000
NICU	500 000	5 000	100/pt.-day		50 000	8 000	75 000
Other	30 500 000				1 700 000	562 000	1 200 000
Subtotal	*35 000 000*				*2 000 000*	*600 000*	*1 300 000*
Hospital total	**50 000 000**				**3 000 000**	**800 000**	**1 500 000**

[a] Direct cost consists of salaries plus supplies. [b] Lab output is in DBS units and NICU output is in patient-days.

Table 4.2. *Method 1—ignore overhead*

Lab cost/DBS unit = \$4 000 000/8 000 000 = \$0.50/DBS unit
NICU cost/pt-day = \$500 000/5000 = \$100/pt-day

Table 4.3. *Method 2—direct allocation of overhead*
(Note: Allocation denominator = sum of 'final' department.)

Lab cost = direct cost + lab's share of admin + lab's share of
    housekeeping + lab's share of laundry

$$= 4\ 000\ 000 + \frac{250\ 000}{2\ 000\ 000}(2\ 000\ 000) + \frac{30\ 000}{600\ 000}(1\ 500\ 000) +$$

$$\frac{25\ 000}{1\ 300\ 000}(1\ 300\ 000)$$

$$= 4\ 000\ 000 + 250\ 000 + 75\ 000 + 25\ 000 = 4\ 350\ 000$$

Lab cost/DBS unit = 4 350 000/8 000 000 = \$0.54/DBS unit

NICU cost = direct cost + share of admin + share of housekeeping +
    share of laundry

$$= 500\ 000 + \frac{50\ 000}{2\ 000\ 000}(2\ 000\ 000) + \frac{8\ 000}{600\ 000}(1\ 500\ 000) +$$

$$\frac{75\ 000}{1\ 300\ 000}(1\ 300\ 000)$$

$$= 500\ 000 + 50\ 000 + 20\ 000 + 75\ 000 + \$645\ 000$$

NICU cost/pt-day = 645 000/5000 = \$129/pt-day

Table 4.4. *Method 3—Step down allocation of overhead*
(Note: Allocation denominator = sum of remaining departments in the step down sequence.)

	Admin	HK	Laundry	Other	Lab	NICU	Other	√
Direct cost	2 000 000	1 500 000	1 300 000	10 200 000	4 000 000	500 000	30 500 000	50m
Allocate Admin	2 000 000 →	$\frac{3}{28}$ = 214 286	$\frac{2}{28}$ = 142 857	$\frac{3}{28}$ = 214 286	$\frac{2.5}{28}$ = 178 571	$\frac{0.5}{28}$ = 35 714	$\frac{17}{28}$ = 1 214 286	
Allocate HK		1 714 286 →	$\frac{8}{766}$ = 17 904	$\frac{158}{766}$ = 353 599	$\frac{30}{766}$ = 67 139	$\frac{8}{766}$ = 17 904	$\frac{562}{766}$ = 1 257 740	
Allocate Laundry			1 460 761 →	$\frac{120}{1420}$ = 123 445	$\frac{25}{1420}$ = 25 718	$\frac{75}{1420}$ = 77 153	$\frac{1200}{1420}$ = 1 234 446	
Total cost				10 891 330	4 271 428	630 771	34 206 472	50m
Units					÷ 8 000 000	÷ 5 000		
Cost/unit					$0.53/DBS unit	$126.15/pt-day		

Table 4.5 *Method 4—Step down with iterations*
(Note: Allocation denominator = sum of all departments except the one being allocated.)

	Admin	HK	Laundry	Other	Lab	NICU	Other	√
Iteration 1								
Direct cost	2 000 000	1 500 000	1 300 000	10 200 000	4 000 000	500 000	30 500 000	50m
Allocate Admin	2 000 000 →	$\frac{3}{28}$ = 214 286	$\frac{2}{28}$ = 142 857	$\frac{3}{28}$ = 214 286	$\frac{2.5}{28}$ = 178 571	$\frac{0.5}{28}$ = 35 714	$\frac{17}{28}$ = 1 214 286	
Allocate HK	$\frac{30}{796}$ = 64 609	→ 1 714 286 →	$\frac{8}{796}$ = 17 229	$\frac{158}{796}$ = 340 273	$\frac{30}{796}$ = 64 609	$\frac{8}{796}$ = 17 229	$\frac{562}{796}$ = 1 210 338	
Allocate Laundry	$\frac{0}{1500}$ = 0	$\frac{80}{1500}$ = 77 871	↓ 1 460 086 →	$\frac{120}{1500}$ = 116 807	$\frac{25}{1500}$ = 24 335	$\frac{75}{1500}$ = 73 004	$\frac{1200}{1500}$ = 1 168 069	
New totals	64 609	77 871	0	10 871 366	4 267 515	625 947	34 092 693	50m
Iteration 2								
Allocate Admin	64 609 →	$\frac{3}{28}$ = 6 922	$\frac{2}{28}$ = 4 615	$\frac{3}{28}$ = 6 922	$\frac{2.5}{28}$ = 5 769	$\frac{0.5}{28}$ = 1 154	$\frac{17}{28}$ = 39 227	
Allocate HK	$\frac{30}{796}$ = 3 196	→ 84 793 →	$\frac{8}{796}$ = 852	$\frac{158}{796}$ = 16 831	$\frac{30}{796}$ = 3 196	$\frac{8}{796}$ = 852	$\frac{562}{796}$ = 59 866	
Allocate Laundry	$\frac{0}{1500}$ = 0	$\frac{80}{1500}$ = 292	↓ 5 467 →	$\frac{120}{1500}$ = 437	$\frac{25}{1500}$ = 91	$\frac{75}{1500}$ = 273	$\frac{1200}{1500}$ = 4 374	

New totals	3 196	292	0	10 895 556	4 276 571	628 226	34 196 160	50m
Iteration 3								
Allocate Admin	3 196 →	$\frac{3}{28}$ = 342	$\frac{2}{28}$ = 228	$\frac{3}{28}$ = 342	$\frac{2.5}{28}$ = 285	$\frac{0.5}{28}$ = 57	$\frac{17}{28}$ = 1 940	
Allocate HK	$\frac{30}{796}$ = 24	634 →	$\frac{8}{796}$ = 6 →	$\frac{158}{796}$ = 126	$\frac{30}{796}$ = 24	$\frac{8}{796}$ = 6	$\frac{562}{796}$ = 448	
Allocate Laundry	$\frac{0}{1500}$ = 0	$\frac{80}{1500}$ = 12	234 →	$\frac{120}{1500}$ = 19	$\frac{25}{1500}$ = 4	$\frac{75}{1500}$ = 12	$\frac{1200}{1500}$ = 187	
New totals	24	12	0	10 896 043	4 276 884	628 301	34 198 735	50m
Final Direct Allocations	24 →	12 →		$\frac{3}{23}$ = 3 $\frac{158}{758}$ = 3	$\frac{2.5}{23}$ = 3 $\frac{30}{758}$ = 0	$\frac{0.5}{23}$ = 1 $\frac{8}{758}$ = 0	$\frac{17}{23}$ = 18 $\frac{562}{758}$ = 9	
Final totals	≈0	0	0	10 896 049	4 276 887	628 302	34 198 762	50m
Units	0			0	÷ 8 000 000	÷ 5 000		
Cost/unit					$0.53/DBS unit	$125.66/pt-day		

Table 4.6. *Method 5—simultaneous allocation (reciprocal method)*
(Note: Allocation denominator = sum of all departments).

Admin $\quad C_1 = 2\,000\,000 + \dfrac{2}{30}C_1 + \dfrac{30}{800}C_2$

HK $\quad C_2 = 1\,500\,000 + \dfrac{3}{30}C_1 + \dfrac{4}{800}C_2 + \dfrac{80}{1500}C_3$

Laundry $\quad C_3 = 1\,300\,000 + \dfrac{2}{30}C_1 + \dfrac{8}{800}C_2$

Lab $\quad C_4 = 4\,000\,000 + \dfrac{2.5}{30}C_1 + \dfrac{30}{800}C_2 + \dfrac{25}{1500}C_3$

NICU $\quad C_5 = 500\,000 + \dfrac{0.5}{30}C_1 + \dfrac{8}{800}C_2 + \dfrac{75}{1500}C_3$

$$\dfrac{28}{30}C_1 - \dfrac{30}{800}C_2 \qquad\qquad\qquad = 2\,000\,000$$

$$-\dfrac{3}{30}C_1 + \dfrac{796}{800}C_2 - \dfrac{80}{1500}C_3 \qquad = 1\,500\,000$$

$$-\dfrac{2}{30}C_1 - \dfrac{8}{800}C_2 \qquad + C_3 \qquad = 1\,300\,000$$

$$-\dfrac{2.5}{30}C_1 - \dfrac{30}{800}C_2 - \dfrac{25}{1500}C_3 + C_4 = 4\,000\,000$$

$$-\dfrac{0.5}{30}C_1 - \dfrac{8}{800}C_2 - \dfrac{75}{1500}C_3 + C_5 = 500\,000$$

The solution of this set of equations is:

$C_1 = 2\,215\,531$
$C_2 = 1\,808\,772$
$C_3 = 1\,465\,790$
$C_4 = 4\,276\,886$
$C_5 = 628\,303$

Therefore, the Cost/unit of output is:

Lab: $\qquad\quad$ $\$4\,276\,886/8\,000\,000 = \underline{\$0.53/\text{DBS unit}}$

NICU: \$$628\,303/5\,000 \qquad = \underline{\$125.66/\text{pt-day}}$

Table 4.7. *Method 6—patient-day allocation of overhead*

This is the simple method described in the footnote of page 27 of Chapter 3 and on page 46 of Chapter 4. It may be useful in some cases.

Laboratory costs would be charged without overhead: $0.50/DBS unit.

NICU costs would be the direct costs of $500 000 plus a share of all relevant other departments (2.0m + 1.5m + 1.3m = 4.8m) in proportion to patient-days (5 000/500 000 where the denominator is total annual hospital patient-days). Thus,

NICU cost = $500 000 + $4 800 000 (5 000/500 000) = $548 000.

NICU cost/pt-day = $548 000/5 000 = $110/pt-day.

4.4. CRITICAL APPRAISAL OF A PUBLISHED ARTICLE

Reference: Lowson, K. V., Drummond,M. F., and Bishop, J. M. (1981). Costing new services: Long-term domiciliary oxygen therapy. *Lancet* **i**, 1146–9.

This paper is assessed below using the 10 questions set out in Annex 3.1. It is suggested that you locate the article and attempt the exercise before reading the assessment.

1. Was a well defined question posed in answerable form?

  x   YES   NO   CAN'T TELL

The study addresses the question, *What is the most efficient way to administer long-term oxygen therapy?* The analysis proceeds by comparing the costs of alternative methods of providing oxygen therapy. However, the effects of the therapy are not included for consideration. The rationale for the omission is that the effects generated by each option are similar, and thus a cost analysis is the most appropriate study design. The extent to which this is true depends, of course, on the accuracy of the assumption of similar effects.

The alternatives compared are cylinder oxygen (large and small containers), liquid oxygen and oxygen concentrators (a machine that extracts oxygen from air).

Cost analysis

The viewpoint for the analysis is not specified explicitly. However, the costs chosen for inclusion suggest that the authors adopt the combined viewpoint of both the United Kingdom National Health Service (NHS) and patients.

2. Was a comprehensive description of the competing alternatives given?

__x__ YES   _____ NO   _____ CAN'T TELL

The three alternatives are fairly well described under the section *Methods of Oxygen Administration*. The paper explicitly excludes the *do-nothing* option, as the question of whether treatment is worthwhile *per se* is not being addressed (and presumably the effects associated with *no treatment* preclude it from being a plausible alternative). The paper alludes to, but does not describe in detail, the alternative of renting concentrators. Since this option is fairly new and untried, the omission is acceptable. Moreover, the authors do identify the conditions under which this option might become attractive in the future.

3. Was there evidence that the programmes' effectiveness had been established?

_____ YES   _____ NO   __x__ CAN'T TELL

The authors cite two studies, references 7 and 8 in the paper, which have confirmed that long-term domiciliary oxygen therapy reduces mortality and improves the quality of life. The relative effectiveness of each alternative was assessed in the second study specifically. The quality of evidence, however, cannot be determined without first consulting the supporting references. Nevertheless, in terms of the classification of forms of analysis outlined in Chapter 2, the study probably qualifies as a *cost-minimization analysis*.

4. Were all the important and relevant costs and consequences for each alternative identified?

_____ YES   _____ NO   __x__ CAN'T TELL

The costs considered in the study are those falling on the NHS and patients, as a result of the incremental resource use of oxygen therapy (i.e., only those treatment costs which are different across treatment options). But the cost estimates, as currently presented, do not allow separate identification of the costs to either the NHS or patients. (For example, there is no way to determine how much patients would have to pay for necessary items such as electricity for running concentrators).

The costs which are common to all options (e.g., physician visits, hospitalization, etc.) are excluded. This approach to costing is appropriate, as long as the actual amounts of the 'common' services consumed are indeed identical for all alternatives. Although costing only the differences in resource use brought about by the alternatives may facilitate the costing exercise, readers should be aware that if this approach is employed, nothing can be concluded about the total cost (i.e., total resource commitment) of any alternative.

The authors choose therapeutic effects as the outcome measure. But what if one option, although producing therapeutic effects equal in magnitude to the other options, is inherently more attractive because it allows, say, greater mobility for the patient? The authors contend that all options *improve the quality of life*, but the relative contribution of each to increasing patients' utility is not known.

5. Were costs and consequences measured accurately in appropriate physical units?

_____ YES _____ NO ___x___ CAN'T TELL

It is difficult to determine whether costs have been measured accurately because only total costs are provided for items such as running costs, rather than the individual quantities and prices of component costs.

An interesting point concerning the relationship between total costs and joint costs is illustrated in Table 4 of the paper. Method A assumes a new workshop facility must be built to maintain the concentrators effectively, while Method B assumes existing facilities are capable of accommodating the workshop. Therefore, with Method A all of the workshop set-up costs and running costs are attributed to the introduction of the concentrators; with Method B, though, many of these costs would have been incurred even without the introduction of concentrators. Consequently, only a portion of the running costs of the whole workshop need to be apportioned to the concentrator servicing, which accounts for the difference in costs between the two methods.

Cost analysis

6. Were costs and consequences valued credibly?

__x__ YES	__x__ NO	_____ CAN'T TELL
(for costs)	(for consequences)	

Market prices were used to estimate costs, based on a typical pattern of care (15 hours oxygen therapy). There is no reason to believe that in this study market prices do not adequately reflect the true cost of resources used. Given the concentration on costs, however, there is little discussion of the valuation of consequences. The authors state that the therapy reduces mortality and improves the general quality of life, and that the alternative modes of delivery *all seemed to be equally effective*. But it is impossible to determine the outcome measures used to derive this conclusion. A more sophisticated *cost–utility analysis* design would have employed a more systematic approach to the measurement of quality of life.

7. Were costs and consequences adjusted for differential timing?

__x__ YES _____ NO _____ CAN'T TELL

Adjustments are made to allow for the fact that, under the concentrator option, more resource outlays will occur earlier as capital expenditures are required for the machines and workshop facilities. This is done by converting the initial lump-sum purchase price into an equivalent annual cost using a seven per cent discount rate. Although no theoretical rationale for the choice is provided, 7 per cent was the public sector discount rate advised by the UK Treasury at the time of the study. (This reinforces the notion that the viewpoint is primarily that of the NHS.)

8. Was an incremental analysis of costs and consequences of alternatives performed?

__x__ YES _____ NO _____ CAN'T TELL

A graph is used to illustrate the differential costs between alternatives. (Recall that the difference in effects is assumed to be zero.) This type of graphical presentation is especially useful when costs vary as the service(s) is(are) expanded. The difference in distance between any two lines can be interpreted as the *incremental cost* of one option over the other.

Readers should not confuse the concepts of *incremental cost* and *marginal cost*. The *incremental cost* is the additional cost that one service or programme imposes over another; however, *marginal cost* is the change in total cost resulting from a one-unit expansion or contraction of the service programme. Thus *incremental cost* is concerned with cost differences between services or programmes, while *marginal cost* is concerned with cost differences within services or programmes. Figure 1 shows that as the number of patients increases, the average costs for the concentrator options fall. (Average costs for the other options remain constant). Figure 2, however, shows that as the number of patients increases, the *incremental costs* between the concentrator options and the other options actually increase (i.e., the distances between the lines get bigger). This is a situation in which, as the number of patients expands, falling average costs make the concentrator option increasingly more cost-effective.

9. Was a sensitivity analysis performed?

_____ YES _____ NO ___x___ CAN'T TELL

The sensitivity of the results to (1) variations in the number of patients serviced, (2) variations in the discount rate, (3) alternative assumptions about the existence of workshop facilities and (4) the length of useful life of the capital cost items was tested. No justification is provided for the range of patients considered (0–100) and the authors do not mention which number is likely to be realistic for practical purposes. The range of discount rates chosen for the sensitivity analysis is not provided, and the results are summarized with the sole comment, 'The choice of discount rate has little effect on the results'. (Perhaps this brevity is a result of space restrictions imposed by the journal.) The rationale for the assumptions about the existence of workshop facilities is well described on p. 1147, column 2. Figure 2 indicates that the results are more sensitive to the choice of assumption when the number of patients serviced is low. As the number of patients increases, the cost difference between the two methods becomes less pronounced.

There is no indication as to the range of assumptions used concerning the length of life of capital equipment. The authors claim, albeit without evidence , that, '. . . the assumption of longer lives does not change the results very much'.

Apart from the capital cost items discussed above, the various cost components were not subjected to any sensitivity testing. The authors should have explored the extent to which different assumptions about these would change the study results.

Cost analysis

10. Did the presentation and discussion of study results include all issues of concern to users?

___x___ YES _____ NO _____ CAN'T TELL

The study compares, in a graphical format, the total costs of providing oxygen for all methods and concludes that, for all but the smallest numbers of patients, concentrators are the most economically efficient methods of service delivery.

The graphical presentation has certain advantages for an evaluation such as this, where the costs of the alternatives are a function of the scale of operation. Figure 1, for example, is a good illustration of how programmes or services which are characterized by high set-up costs may initially appear to be relatively more costly than other options, but when costs are spread over an increasingly greater number of patients, the average cost per patient falls.

A significant implication of this finding is that the average cost per patient for the concentrator option will depend on the actual number of patients receiving care. If a concentrator programme is running with fewer than 13 patients, the other alternatives will prove to be more efficient. The authors recognize this point and caution, 'What is cost-effective in one location may not necessarily be cost-effective in another'.

The authors cite two barriers to possible implementation of the concentrator methods. First, the concentrators require a large capital outlay which might exceed budgetary allowances. This problem can be at least partially overcome, it is suggested, by introducing more flexibility into the budgetary allocation process. Second, the provision of concentrators would fall on a different budgetary authority from the one currently funding small cylinder provision, thereby creating a disincentive for adopting the less costly concentrator method. Administrative change is recommended to overcome this obstacle.

REFERENCES

Boyle, M. H., Torrance, G. W., Horwood, S. P., Sinclair, J. C. (1982). *A cost analysis of providing neonatal intensive care to 500–1499 gram birth weight infants*. Research Report #51, Programme for Quantitative Studies in Economics and Population, McMaster University, Hamilton, Canada.

Clements, R. M. (1974). *The Canadian hospital accounting manual supplement*. Livingston Printing, Toronto.

Horngren, C. T. (1982). *Cost accounting: a managerial emphasis* (5th edn). Prentice Hall, Englewood Cliffs, N. J.

Hull, R., Hirsh, J., Sackett, D. L., and Stoddart, G. L. (1982). Cost-effectiveness of primary and secondary prevention of fatal pulmonary embolism in high-risk surgical patients. *Can. Med. Assoc. J.* **127**, 990–5.

Kaplan, R. S. (1973). Variable and self-service costs in reciprocal allocation models. *The Accounting Review* **XLVIII**, 738–48.

Levin, H. M. (1975). Cost-effectiveness analysis in evaluation research. In M. Guttentag and E. L. Struening (eds) *Handbook of evaluation research*, Vol. 2 pp. 89–122. Sage, London.

Lowson, K. V., Drummond, M. F., and Bishop, J. M. (1981). Costing new services: Long-term domiciliary oxygen therapy. *Lancet* **i**, 1146–9.

Neuhauser, D. and Lewicki, A. M. (1975). What do we gain from the sixth stool guaiac? *N. Engl. J. Med.* **293**(5), 226–8.

Richardson, A. W. and Gafni, A. (1983). Treatment of capital costs in evaluating health care programmes. *Cost and Management* **Nov–Dec:** 26–30.

ANNEX 4.1. METHODS OF MEASURING AND VALUING CAPITAL COSTS

We are indebted to Morris Barer of the University of British Columbia for producing these examples, which should clarify the treatment of capital costs.

As a first note, we need to distinguish two classes of 'capital'—land and equipment. This is an important consideration, because in costing exercises we assume land does not depreciate, while of course capital equipment does. You can think of there being a continuum along which materials and supplies 'depreciate' or are used up instantaneously and so are costed fully in the year of use; capital equipment depreciates more slowly, and may be handled in a variety of ways; land does not depreciate at all.

As a second note, recall that 'capital equipment costs have three components—depreciation, opportunity cost, and actual operating costs. We will ignore the last of these here.

First consider *equipment*, and let us use an example of a machine costing $200 000 that, at the end of 5 years, has re-sale value of $20 000. Assume straight-line depreciation and a discount rate of 4 per cent. There are, then, four approaches to costing:

(i) one can assume all costs accrue at time 0. This amounts to treating the equipment as one would less durable materials and supplies:

Cost analysis

Time	0	1	2	3	4	5
Depreciation	200 000	0	0	0	0	(20 000)
Undepreciated balance at beginning of period	—	0	0	0	0	0
Opportunity cost	—	0	0	0	0	0
Dep'n. + opp cost	200 000	0	0	0	0	(20 000)
Present value (PV)	200 000	0	0	0	0	(16 439)

Net present value (NPV) of equipment cost = $183 561

Alternatively, but equivalently, one can treat the machine as instantaneously depreciating, except for the $20 000 resale value, which then is maintained through the 5 years:

Time	0	1	2	3	4	5
Depreciation	180 000	—	—	—	—	—
Undepreciated balance at beginning of period		20 000	20 000	20 000	20 000	20 000
Opportunity cost		800	800	800	800	800
Dep'n. + opp cost	180 000	800	800	800	800	800
PV	180 000	769	740	711	684	658

NPV of equipment cost = $183 562

(ii) One can compute depreciation and opportunity costs separately. They are related in that the opportunity cost of equipment refers to the use of the rsources embodied in the equipment, in their next best use—this is 'approximated' by calculating the return on the funds implicit in the undepreciated value of the equipment at each point in time. Hence, the higher the rate of depreciation, the lower the opportunity cost, all else equal. Again, one has the choice of building the $20 000 resale in at the end, or just depreciating less of the machine. It works out the same:

Time	1	2	3	4	5
Depreciation	36 000	36 000	36 000	36 000	36 000
Undepreciated balance at beginning of period	200 000	164 000	128 000	92 000	56 000
Opportunity cost	8 000	6 560	5 120	3 680	2 240
Dep'n. + opp cost	44 000	42 560	41 120	39 680	38 240
PV	42 308	39 349	36 556	33 919	31 430

NPV of equipment cost = $183 562

Time	1	2	3	4	5
Depreciation	40 000	40 000	40 000	40 000	20 000
Undepreciated balance at beginning of period	200 000	160 000	120 000	80 000	40 000
Opportunity cost	8 000	6 400	4 800	3 200	1 600
Dep'n. + opp cost	48 000	46 400	44 800	43 200	21 600
PV	46 154	42 899	39 827	36 928	17 754

NPV of equipment cost = $183 562

(iii) One can compute an equivalent annual cost. This may be useful in a situation where other operating costs are the same each year, making necessary the comparison of only a single year of cost data for each alternative in the economic evaluation:

NPV $= E \cdot AF_{5,4\%}$ (Where $AF_{5,4\%}$ is the annuity factor for 5 years at an interest rate of 4 per cent. See Table 2 in Annex 4.2)

$183\ 562 = E \cdot 4.4518 \rightarrow E = \$41\ 233$

In other words, an *equal* stream of costs amounting to $41 233 in *each* of the five years of the program has a present value equivalent to any of the *unequal* cost streams in (i) or (ii) above. Note, therefore, that the equivalent annual cost embodies both depreciation and opportunity cost.

Cost Analysis

(iv) One can use equivalent or actual rental costs, if available or estimable. Note that because the renter will need to recover not only depreciation of the rental equipment but also a rate of return at least as good as that from the next best use of the resource, one can take rental cost to embody both depreciation and opportunity cost.

Second, the treatment of *land* is quite different because of the lack of depreciation. A land purchase of $200 000 at time 0 would generate the following cost time stream:

Time	1	2	3	4	5
Depreciation	—	—	—	—	—
Undepreciated balance at beginning of period	200 000	200 000	200 000	200 000	200 000
Opportunity cost	8 000	8 000	8 000	8 000	8 000
Dep'n + opp cost	8 000	8 000	8 000	8 000	8 000
PV	7 692	7 396	7 112	6 838	6 575
NPV = $35 613					

Converted to an equivalent annual cost.

$$NPV = E \cdot AF_{5, 4\%}$$
$$\$35\ 613 = E \cdot 4.4518$$

It comes as no particular surprise that E = $8000!

Annex 4.2. Discount Table 1

Present value of $1

N	1%	2%	3%	4%	5%	6%	7%	8%	9%	10%	11%	12%	13%	14%	15%
1	0.9901	0.9804	0.9709	0.9615	0.9524	0.9434	0.9346	0.9259	0.9174	0.9091	0.9009	0.8929	0.8850	0.8772	0.8696
2	0.9803	0.9612	0.9426	0.9246	0.9070	0.8900	0.8734	0.8573	0.8417	0.8264	0.8116	0.7972	0.7831	0.7695	0.7561
3	0.9706	0.9423	0.9151	0.8890	0.8638	0.8396	0.8163	0.7938	0.7722	0.7513	0.7312	0.7118	0.6931	0.6750	0.6575
4	0.9610	0.9238	0.8885	0.8548	0.8227	0.7921	0.7629	0.7350	0.7084	0.6830	0.6587	0.6355	0.6133	0.5921	0.5718
5	0.9515	0.9057	0.8626	0.8219	0.7835	0.7473	0.7130	0.6806	0.6499	0.6209	0.5935	0.5674	0.5428	0.5194	0.4972
6	0.9420	0.8880	0.8375	0.7903	0.7462	0.7050	0.6663	0.6302	0.5963	0.5645	0.5346	0.5066	0.4803	0.4556	0.4323
7	0.9327	0.8706	0.8131	0.7599	0.7107	0.6651	0.6227	0.5835	0.5470	0.5132	0.4817	0.4523	0.4251	0.3996	0.3759
8	0.9235	0.8535	0.7894	0.7307	0.6768	0.6274	0.5820	0.5403	0.5019	0.4665	0.4339	0.4039	0.3762	0.3506	0.3269
9	0.9143	0.8368	0.7664	0.7026	0.6446	0.5919	0.5439	0.5002	0.4604	0.4241	0.3909	0.3606	0.3329	0.3075	0.2843
10	0.9053	0.8203	0.7441	0.6756	0.6139	0.5584	0.5083	0.4632	0.4224	0.3855	0.3522	0.3220	0.2946	0.2697	0.2472
11	0.8963	0.8043	0.7224	0.6496	0.5847	0.5268	0.4751	0.4289	0.3875	0.3505	0.3173	0.2875	0.2607	0.2366	0.2149
12	0.8874	0.7885	0.7014	0.6246	0.5568	0.4970	0.4440	0.3971	0.3555	0.3186	0.2858	0.2567	0.2307	0.2076	0.1869
13	0.8787	0.7730	0.6810	0.6006	0.5303	0.4688	0.4150	0.3677	0.3262	0.2897	0.2575	0.2292	0.2042	0.1821	0.1625
14	0.8700	0.7579	0.6611	0.5775	0.5051	0.4423	0.3878	0.3405	0.2992	0.2633	0.2320	0.2046	0.1807	0.1597	0.1413
15	0.8613	0.7430	0.6419	0.5553	0.4810	0.4173	0.3624	0.3152	0.2745	0.2394	0.2090	0.1827	0.1599	0.1401	0.1229
16	0.8528	0.7284	0.6232	0.5339	0.4581	0.3936	0.3387	0.2919	0.2519	0.2176	0.1883	0.1631	0.1415	0.1229	0.1069
17	0.8444	0.7142	0.6050	0.5134	0.4363	0.3714	0.3166	0.2703	0.2311	0.1978	0.1696	0.1456	0.1252	0.1078	0.0929
18	0.8360	0.7002	0.5874	0.4936	0.4155	0.3503	0.2959	0.2502	0.2120	0.1799	0.1528	0.1300	0.1108	0.0946	0.0808
19	0.8277	0.6864	0.5703	0.4746	0.3957	0.3305	0.2765	0.2317	0.1945	0.1635	0.1377	0.1161	0.0981	0.0829	0.0703
20	0.8195	0.6730	0.5537	0.4564	0.3769	0.3118	0.2584	0.2145	0.1784	0.1486	0.1240	0.1037	0.0868	0.0728	0.0611
21	0.8114	0.6598	0.5375	0.4388	0.3589	0.2942	0.2415	0.1987	0.1637	0.1351	0.1117	0.0926	0.0768	0.0638	0.0531
22	0.8034	0.6468	0.5219	0.4220	0.3418	0.2775	0.2257	0.1839	0.1502	0.1228	0.1007	0.0826	0.0680	0.0560	0.0462
23	0.7954	0.6342	0.5067	0.4057	0.3256	0.2618	0.2109	0.1703	0.1378	0.1117	0.0907	0.0738	0.0601	0.0491	0.0402
24	0.7876	0.6217	0.4919	0.3901	0.3101	0.2470	0.1971	0.1577	0.1264	0.1015	0.0817	0.0659	0.0532	0.0431	0.0349
25	0.7798	0.6095	0.4776	0.3751	0.2953	0.2330	0.1842	0.1460	0.1160	0.0923	0.0736	0.0588	0.0471	0.0378	0.0304

Annex 4.2. Discount Table 1 (*cont.*)

Present value of $1

N	1%	2%	3%	4%	5%	6%	7%	8%	9%	10%	11%	12%	13%	14%	15%
26	0.7720	0.5976	0.4637	0.3607	0.2812	0.2198	0.1722	0.1352	0.1064	0.0839	0.0663	0.0525	0.0417	0.0331	0.0264
27	0.7644	0.5859	0.4502	0.3468	0.2678	0.2074	0.1609	0.1252	0.0976	0.0763	0.0597	0.0469	0.0369	0.0291	0.0230
28	0.7568	0.5744	0.4371	0.3335	0.2551	0.1956	0.1504	0.1159	0.0895	0.0693	0.0538	0.0419	0.0326	0.0255	0.0200
29	0.7493	0.5631	0.4243	0.3207	0.2429	0.1846	0.1406	0.1073	0.0822	0.0630	0.0485	0.0374	0.0289	0.0224	0.0174
30	0.7419	0.5521	0.4120	0.3083	0.2314	0.1741	0.1314	0.0994	0.0754	0.0573	0.0437	0.0334	0.0256	0.0196	0.0151
35	0.7059	0.5000	0.3554	0.2534	0.1813	0.1301	0.0937	0.0676	0.0490	0.0356	0.0259	0.0189	0.0139	0.0102	0.0075
40	0.6717	0.4529	0.3066	0.2083	0.1420	0.0972	0.0668	0.0460	0.0318	0.0221	0.0154	0.0107	0.0075	0.0053	0.0037
45	0.6391	0.4102	0.2644	0.1712	0.1113	0.0727	0.0476	0.0313	0.0207	0.0137	0.0091	0.0061	0.0041	0.0027	0.0019
50	0.6080	0.3715	0.2281	0.1407	0.0872	0.0543	0.0339	0.0213	0.0134	0.0085	0.0054	0.0035	0.0022	0.0014	0.0009

Annex 4.2. Discount Table 2

Present value of annuity of $1 in arrears

N	1%	2%	3%	4%	5%	6%	7%	8%	9%	10%	11%	12%	13%	14%	15%
1	0.9901	0.9804	0.9709	0.9615	0.9524	0.9434	0.9346	0.9259	0.9174	0.9091	0.9009	0.8929	0.8850	0.8772	0.8696
2	1.9704	1.9416	1.9135	1.8861	1.8594	1.8334	1.8080	1.7833	1.7591	1.7335	1.7125	1.6901	1.6681	1.6467	1.6257
3	2.9410	2.8839	2.8286	2.7751	2.7232	2.6730	2.6243	2.5771	2.5313	2.4869	2.4437	2.4018	2.3612	2.3216	2.2832
4	3.9020	3.8077	3.7171	3.6299	3.5460	3.4651	3.3872	3.3121	3.2397	3.1699	3.1024	3.0373	2.9745	2.9137	2.8550
5	4.8534	4.7135	4.5797	4.4518	4.3295	4.2124	4.1002	3.9927	3.8897	3.7908	3.6959	3.6048	3.5172	3.4331	3.3522
6	5.7955	5.6014	5.4172	5.2421	5.0757	4.9173	4.7665	4.6229	4.4859	4.3553	4.2305	4.1114	3.9975	3.8887	3.7845
7	6.7282	6.4720	6.2303	6.0021	5.7864	5.5824	5.3893	5.2064	5.0330	4.8684	4.7122	4.5638	4.4226	4.2883	4.1604
8	7.6517	7.3255	7.0197	6.7327	6.4632	6.2098	5.9713	5.7466	5.5348	5.3349	5.1461	4.9676	4.7988	4.6389	4.4873
9	8.5660	8.1622	7.7861	7.4353	7.1078	6.8017	6.5152	6.2469	5.9952	5.7590	5.5370	5.3282	5.1317	4.9464	4.7716
10	9.4713	8.9826	8.5302	8.1109	7.7217	7.3601	7.0236	6.7101	6.4177	6.1446	5.8892	5.6502	5.4262	5.2161	5.0188

12	11.2551	10.5753	9.9540	9.3851	8.8633	8.3838	7.9427	7.5361	7.1607	6.8137	6.4924	6.1944	5.9176	5.6603	5.4206
13	12.1337	11.3484	10.6350	9.9856	9.3936	8.8527	8.3577	7.9038	7.4869	7.1034	6.7499	6.4235	6.1218	5.8424	5.5831
14	13.0037	12.1062	11.2961	10.5631	9.8986	9.2950	8.7455	8.2442	7.7862	7.3667	6.9819	6.6282	6.3025	6.0021	5.7245
15	13.8651	12.8493	11.9379	11.1184	10.3797	9.7122	9.1079	8.5595	8.0607	7.6061	7.1909	6.8109	6.4624	6.1422	5.8474
16	14.7179	13.5777	12.5611	11.6523	10.8378	10.1059	9.4466	8.8514	8.3126	7.8237	7.3792	6.9740	6.6039	6.2651	5.9542
17	15.5623	14.2919	13.1661	12.1657	11.2741	10.4773	9.7632	9.1216	8.5436	8.0216	7.5488	7.1196	6.7291	6.3729	6.0472
18	16.3983	14.9920	13.7535	12.6593	11.6896	10.8276	10.0591	9.3719	8.7556	8.2014	7.7016	7.2497	6.8399	6.4674	6.1280
19	17.2260	15.6785	14.3238	13.1339	12.0853	11.1581	10.3356	9.6036	8.9501	8.3649	7.8393	7.3658	6.9380	6.5504	6.1982
20	18.0456	16.3514	14.8775	13.5903	12.4622	11.4699	10.5940	9.8181	9.1285	8.5136	7.9633	7.4694	7.0248	6.6231	6.2593
21	18.8570	17.0112	15.4150	14.0292	12.8212	11.7641	10.8355	10.0168	9.2922	8.6487	8.0751	7.5620	7.1016	6.6870	6.3125
22	19.6604	17.6580	15.9369	14.4511	13.1630	12.0416	11.0612	10.2007	9.4424	8.7715	8.1757	7.6446	7.1695	6.7429	6.3587
23	20.4558	18.2922	16.4436	14.8568	13.4886	12.3034	11.2722	10.3711	9.5802	8.8832	8.2664	7.7184	7.2297	6.7921	6.3988
24	21.2434	18.9139	16.9355	15.2470	13.7986	12.5504	11.4693	10.5288	9.7066	8.9847	8.3481	7.7843	7.2829	6.8351	6.4338
25	22.0232	19.5235	17.4131	15.6221	14.0939	12.7834	11.6536	10.6748	9.8226	9.0770	8.4217	7.8431	7.3300	6.8729	6.4641
26	22.7952	20.1210	17.8768	15.9828	14.3752	13.0032	11.8258	10.8100	9.9290	9.1609	8.4881	7.8957	7.3717	6.9061	6.4906
27	23.5596	20.7069	18.3270	16.3296	14.6430	13.2105	11.9867	10.9352	10.0266	9.2372	8.5478	7.9426	7.4086	6.9352	6.5135
28	24.3164	21.2813	18.7641	16.6631	14.8981	13.4062	12.1371	11.0511	10.1161	9.3066	8.6016	7.9844	7.4412	6.9607	6.5335
29	25.0658	21.8444	19.1885	16.9837	15.1411	13.5907	12.2777	11.1584	10.1983	9.3696	8.6501	8.0218	7.4701	6.9830	6.5509
30	25.8077	22.3965	19.6004	17.2920	15.3725	13.7648	12.4090	11.2578	10.2737	9.4269	8.6938	8.0552	7.4957	7.0027	6.5660
35	29.4086	24.9986	21.4872	18.6646	16.3742	14.4982	12.9477	11.6546	10.5668	9.6442	8.8552	8.1755	7.5856	7.0700	6.6166
40	32.8347	27.3555	23.1148	19.7928	17.1591	15.0463	13.3317	11.9246	10.7574	9.7791	8.9511	8.2438	7.6344	7.1050	6.6418
45	36.0945	29.4902	24.5187	20.7200	17.7741	15.4558	13.6055	12.1084	10.8812	9.8628	9.0079	8.2825	7.6690	7.1232	6.6543
50	39.1961	31.4236	25.7298	21.4822	18.2559	15.7619	13.8007	12.2335	10.9617	9.9148	9.0417	8.3045	7.6752	7.1327	6.6605

5. Cost-effectiveness analysis

5.1. SOME BASICS

Cost-effectiveness analysis (CEA) is one form of full economic evaluation, where both the costs and consequences of health programmes or treatments are examined. Therefore all the points discussed in Chapter 4 on cost analysis apply here also. This chapter introduces some additional issues that need to be confronted when undertaking a CEA; in addition to the general introduction, it contains a study design exercise (in Section 5.2), followed (in Section 5.3) by a critical appraisal exercise. A particularly important methodological issue, the incremental analysis of costs and effects, is discussed in Section 5.4, and finally, another critical appraisal exercise is given in Section 5.5. As before, the chapter progresses by attempting to answer some of the questions that analysts might need to consider when undertaking a CEA.

5.1.1. What will be the chosen measure of effectiveness?

In general terms the answer to this question lies in the objectives of the programmes or treatments being evaluated, and it is always worthwhile taking time to clarify what these are. Sometimes the objectives will be unclear, often there will be multiple objectives. In order to carry out a cost-effectiveness analysis, one or other of the following conditions must hold:

(a) that there is one, unambiguous, objective of the intervention(s) and therefore a clear dimension along which effectiveness can be assessed; or

(b) that there are many objectives, but that the alternative interventions are thought to achieve these to the same extent.

An example of the first case would be where two therapies could be compared in terms of their cost per year of life gained, or, say, two screening procedures could be compared in terms of the cost per case found. As was mentioned in Chapter 2, comparisons can be made across a broad range of disparate programmes (e.g., treatments for chronic renal disease or seatbelt legislation) if there is a common effect of interest (e.g., lives saved.

74

An example of the second case would be where, say, two surgical interventions gave similar results in terms of complications and recurrences. A cost-effectiveness study in such an instance would, in the terminology of Chapter 2, be called a cost-minimization analysis. (If it were merely a costing study, carried out on the assumption that the effectiveness of the alternatives was equivalent, but without active consideration of that evidence, it might be termed a cost analysis.)

A cost-minimization analysis can be carried out without ambiguity if it is based on existing (medical) evidence of effectiveness. However, if the effectiveness evidence were to be generated at the same time as the costs, one would not know in advance whether equivalence in effects will be obtained. (Although the null hypothesis of the associated clinical trial might give some clues.) Therefore, one might find that a study beginning as a cost-minimization analysis may require a more sophisticated approach. For example, multiple dimensions of effectiveness may need to be assessed relative to one another. (Those faced with this problem should consult Chapter 6, on cost–utility analysis.) Similarly, it might be thought, in the early stages of a cost-effectiveness analysis, that just one dimension of effectiveness were important, only to find that unforeseen effects are also relevant to the assessment. (For example, two diagnostic tests might be compared in terms of cost per case found, only to discover that the approach with the higher effectiveness in case finding resulted in minor clinical complications.)

Therefore, when beginning a study one can never be completely sure of its final form, particularly if the effectiveness evidence is to be generated at the same time as the costs. As was mentioned in Chapter 2, the distinction between cost-effectiveness analysis and cost–utility analysis often becomes blurred. However, the final form of analysis is not too important at the beginning of a study and the following rules of thumb should be of some help.

1. Always take time to clarify the objectives of the programme or treatment.
2. If one major dimension for the measurement of 'success' is apparent, perform a cost-effectiveness analysis based on this dimension. (Or perform a cost-minimization analysis, if it turns out that the alternatives have equivalent effectiveness on the chosen dimension.)
3. Be on the lookout for other attributes of the alternatives being assessed, even if the medical research design does not consider these formally. Where possible, record the effectiveness of the alternatives as judged on these extra dimensions and be prepared to

mount *ad hoc* surveys to obtain more information (e.g., in a study of day care surgery, where the main clinical endpoint might be *number of complications*, it also may be relevant to undertake a patient satisfaction survey).

4. Keep open the possibility of employing more sophisticated forms of analysis if it turns out that there is more than one appropriate dimension for judging effectiveness. The 'utility' assessments (to be discussed in Chapter 6) can always be undertaken separately. Alternatively it might be necessary only to present an array of the differential achievements, by dimension, of the alternative programmes. These can then be given to the decision-maker, at the programmatic or clinical level, so that he can make his own trade-off between effects. This approach is sometimes called the 'score-card' method. It is described in more detail in Quade (1982) and has been used in a recent study by Baynham, Dent, and Torrance (1984).

There is one further important methodological issue to be addressed in the choice of effectiveness measure; namely, should this always relate to a final health *output* such as *life-years gained*, or can it relate to an intermediate output such as *cases found* or *patients appropriately treated*. Intermediate outputs are admissible, although care must be taken to establish a link between these and a final health output, or to show that the intermediate outputs themselves have some value. For example, correct diagnosis of cases and the consequent confirmation of true negatives can provide reassurance both to the patient and to the doctor, and therefore may have a value in its own right quite apart from the health effects resulting from subsequent treatment. In general though, one should choose an effectiveness measure relating to a final output.

5.1.2. How are the effectiveness data to be obtained?

Although primarily an epidemiological issue, the availability of data on the effectiveness of the programmes or treatments being assessed is crucial to the cost-effectiveness analyst. (In fact many CEAs are criticized more often for the quality of the medical evidence on which they are based, rather than for the subsequent economics.)

A major source of effectiveness data is the existing medical literature. Use of such data raises two issues: quality and relevance. Appraisal of the quality of medical evidence is beyond the scope of this book and the reader should consult the clinical epidemiology rounds produced by the Department of Clinical Epidemiology and Biostatistics, McMaster University (1981). These papers set out a checklist of questions to ask of any published study of diagnostic or therapeutic interventions. Although there are a number of important methodological features of a well-designed clinical study, probably the most important aspect is the random allocation of patients to treatment groups (including a control group). Application of this single test would lead one to have serious reservations about the clinical evidence used in many published economic evaluations.

In judging the relevance of results published in the literature, one would have to consider how close one's own situation is to those where the published clinical studies were conducted. Important factors to consider are the patient caseload, the expertise of medical and other staff, and the existence of backup facilities.

If no good clinical evidence exists, the cost-effectiveness analyst has two options. Either he can proceed by making assumptions about some of the medical parameters, or he can design a study that will generate the effectiveness evidence required. Obviously the latter strategy is to be preferred from a scientific point of view, but the former approach should not be dismissed lightly. For example, there may be a number of practical obstacles to mounting a properly designed clinical study. These include resistance from practitioners, time constraints and costs. Of course none of these obstacles should be regarded as immutable. For example, resistance from practitioners could be overcome by explaining that, until a proper evaluation is carried out, one cannot be sure that one's clinical practices are doing more good than harm. Also, a delay in the decision on the adoption of a new programme or treatment may be justified if this enables good quality medical evidence to be generated. However, there may be situations where the obstacles to mounting a study indicate more rough-and-ready approaches.

In such situations, the cost-effectiveness analyst can proceed by making assumptions about the clinical evidence and undertaking a sensitivity analysis of the economic results to different assumptions. (The basic notion of sensitivity analysis was introduced in Chapter 3. The underlying logic is that if the final result is not sensitive to the estimate used for a given variable, then it is not worth much effort to obtain a more accurate estimate.) It may be that in some cases a CEA based on existing medical evidence, with an appropriate sensitivity analysis, can obviate the need for a costly and time consuming clinical trial. This might be the case in extreme situations where a very small improvement in effectiveness (much smaller than that expected to be observed) would make the new programme or treatment cost-effective, or where even high effectiveness of the new programme (much higher than that observed before in similar programmes) would not make the new programme cost-effective.

Cost-effectiveness analysis

At any rate sensitivity analysis can be used to estimate the minimum level of effectiveness required to make the given programme or treatment more cost-effective than the alternative. Given the importance of sensitivity analysis in all economic evaluations, it is discussed in more detail in Section 5.1.6.

If it is decided to choose the other option, of designing a study to generate the required medical evidence, the normal rules of medical research design would apply. However, two additional points are worth noting. First, it might be worthwhile considering modification of the medical research design to make it more amenable to the associated economic research. In the main this would involve modifying some of the data instruments to capture the relevant economic data. Implicitly these data are indicated in this volume, but the reader may also wish to consult Drummond and Stoddart (1984) where the implications of building economics into clinical trials are discussed more fully.

Second, care should be taken to ensure that the very existence of the clinical study does not cause working practices to deviate too much from what is normal, since the cost-effectiveness study needs to be based on the programme or treatment as it would be carried out on a regular basis. [This was one of the arguments given by Evans and Robinson (1980) for not linking their cost-effectiveness study of day care surgery to a prospective controlled clinical trial.]

5.1.3. Should indirect costs and benefits be included?

As was indicated in Chapter 4, the relevance of this depends on the viewpoint for the analysis. However, even when a societal viewpoint is being adopted, the inclusion of this category of cost or consequence is still contentious. In CEA this issue might arise as follows. Suppose one were evaluating two programmes for mental illness patients. One programme requires institutionalization of the patient for a given period; the other, being a community-based programme using community psychiatric nurses in association with out-patient hospital visits, means that patients can remain in their own homes. (For simplicity, let us assume that the programmes turn out to be equivalent in their medical effectiveness, as assessed by some agreed measure of clinical symptomatology. Furthermore, assume that a survey of patients shows them to be indifferent to the treatment modes, providing they are 'cured'.)

Suppose it turns out that the community care programme has higher direct costs, but that the number of workdays lost by the cohort of patients on the regimen is lower, as many more of them can remain at work. Would it be right to deduct these production gains from the higher direct costs of the community care programme? If so how would the production gains be valued?

One might take the view that the production gains should be included in the analysis, since in principle there is no difference between these resource savings and any of the other labour inputs included in the direct cost estimates. Many analysts would follow this approach. The indirect benefits would be estimated by using the extra earnings of patients on the community care regimen, gross of taxes and benefits (i.e., gross earnings before deductions, plus employer-paid benefits). The logic here is that the gross wage reflects the value of the production at the margin.

Whilst the approach followed above is quite defensible, it gives rise to a number of wider considerations that should be noted. First, the approach assumes that the community loses production if the institutional-based programme removes patients from employment. However, it may be that, given a pool of unemployed labour, the jobs vacated by patients admitted to institutional care could be filled by other members of the community. If this were the case there may be few overall production gains from adopting the community care programme.

Second, it may be that at some later stage the cost-effectiveness estimates obtained in this study are compared with those obtained in other fields of health care, say a community care programme for the mentally handicapped or the elderly. Since the patients benefiting from these programmes are unlikely to be in employment, there is less potential for production gains. This would make the community care programme for mental illness patients seem relatively 'cheap' in terms of net cost, particularly if it were for workers earning high incomes, such as business executives, psychiatrists or, dare we say, economists! Thus, in making a choice on the basis of net cost-effectiveness estimates, the decision-maker may be tacitly accepting priorities different from his stated ones—if these are for the care of the elderly or mentally handicapped.

Obviously, the astute decision-maker is unlikely to make such an error, and such comparisons, across a broad range of programmes, are rarely made solely on the basis of cost-effectiveness estimates. However, the economic analyst should be aware of the possibility. The situation may partly be eased by counting a *day of normal activities lost through institutionalization* as representing the same value to the community, no matter what the category of worker (including those *working at home*). Nevertheless the basic conflict between the different resource value of skilled versus unskilled labour (or the young worker versus the elderly or infirm) and the objective of equity in the provision of care is always present.

5.1.4. Should changes in the costs of other treatments or programmes be included?

Another decision about the boundaries of a given CEA concerns the extent to which the impact on other programme costs should be considered. This is perhaps best illustrated by an example: if a programme of screening and treatment for hypertension means that people previously dying of strokes and heart attacks now live to a ripe old age and then die of cancer, should the costs of the resulting cancer therapy be added to the costs of the hypertension programme?

First, it has to be remembered that all the forms of economic evaluation discussed in this book are what economists call partial equilibrium analyses. That is, whilst it is recognized that any change in economic activity (such as investment in health programmes) induces many ripples throughout the economy, it is argued that such investments can be assessed against a background of all else remaining constant. Therefore, an artificial boundary is always being drawn around analyses.

The broader issue apart, the answer to the question posed above depends on the strength of the linkage between the various decisions concerned. For example, if hypertension therapy does extend the lives of people, there is nothing to say that they have to be given cancer therapy at a later stage. This is a decision that should be made on its own merits, depending on the additional costs incurred and the consequences resulting. Therefore, why make the hypertension programme seem less attractive (by adding these extra costs) when the decision to institute cancer therapy can be taken separately? However, in the calculation of the life extension from instituting the hypertension screening programme, if it has been assumed that those developing cancer will have the benefits of life extension from the therapies available, then consistency would demand that the costs of cancer therapy be also included.

The saving grace here (from the point of view of the analyst) is that discounting future costs and effects to present values will reduce their quantitative importance in the analysis. Therefore it may be safe to ignore them in many situations because, in the words of one analyst, they may amount to no more than a 'hill of beans' (Bush 1973).

Nevertheless there may be some instances where the linkages between the programmes being evaluated and subsequent actions are much stronger. For example, there is rarely much point in screening for disease without at the same time intending to administer therapy to the cases found. Therefore, it would usually make sense to treat these decisions as a package with respect to the analysis of their costs and consequences. However, there are also situations where the appropriate treatment of the

linkage between decisions and the boundaries of any economic evaluation, are more perplexing. A good example is the issue of whether, in an
economic evaluation of screening for spina bifida cystica or for Down's
syndrome, one should argue that there is a link between abortion of an
affected foetus and the parents' subsequent decision on replacement of
the pregnancy. [For an insight into the methodological problems that can
arise here see Hagard and Carter (1976) and Henderson (1982).]

5.1.5. Should effects occurring in the future be discounted?

In Chapter 4, the logic for, and the procedures of, discounting costs to
present values were outlined. Since cost-effectiveness analysis also
considers effects, should these be discounted too? This issue has aroused
controversy, although it should be pointed out that in many CEAs it does
not arise because the effects occur in a short period of time. Usually the
costs also occur in a short period of time, although capital costs may have
to be converted to an annual amount using the annuitization procedure
outlined in Chapter 4.

The reasons often given for not discounting effects are that:

(a) unlike resources, it is difficult to conceive of individuals investing
  in health or trading flows of healthy years through time;
(b) discounting years of life gained in the future gives less weight to
  future generations in favour of the present one. Whereas this may
  make sense in the context of resources, where one would expect
  future generations to be wealthier, it might not make sense in the
  context of health. (On the other hand, it might, if one expects future
  generations to have better therapeutic technologies available.)

However, there are some fairly powerful arguments in favour of
discounting effects as well as costs. They are:

(a) it can be shown, by the use of simple numerical examples, that
  leaving effects undiscounted while discounting costs can lead to
  inconsistencies in reasoning. For a simple numerical example see
  Weinstein and Stason (1977); for a theoretical treatment see
  Keeler and Cretin (1983);
(b) leaving effects undiscounted leads to quite impossible conclusions.
  For example, a health programme giving rise to $1 of health
  benefits each and every year stretching into the future would be
  wortwhile whatever the size of the initial capital sum;
(c) contrary to the argument set out above, one *can* conceive of investments in health and the trading of health through time (Grossman

1972). Although it is not possible to give up a year now in return for a year at the end of one's life, individuals can trade reductions in health status or other goods and services now, in return for healthy time in the future (and *vice versa*). If this were not the case, people would not abstain from pleasurable but potentially unhealthy (in the long term) pursuits.

Therefore, the weight of the argument is for discounting health effects occuring in the future, although both theoretical and empirical research is currently being carried out into the rate at which individuals discount health. However, the current state of the art would suggest that effects should be treated in the same way as costs, and discounted at the same rate.

5.1.6. What are the main points to consider when undertaking a sensitivity analysis?

A few years ago, few economic evaluations in the health care field included a comprehensive sensitivity analysis. Nowadays sensitivity analysis is virtually mandatory, and it is likely that the state of the art will improve rapidly now that more analysts with a statistical background are taking an interest in health care evaluation. Therefore, as well as noting the basic hints given below, the reader might be wise to examine practice in the most recently published papers in this field.

At a very basic level, the steps one might take are as follows:

1. Consider which of the estimates made in the analysis are
   (a) subject to debate because no estimates were available and informed guesses were made (e.g., the effectiveness of new, and unproven, medical procedures);
   (b) subject to debate because of known imprecision in the estimation procedure (e.g., hospitalization costs based on average, *per diem*, figures);
   (c) subject to debate because of methodological controversy or the potential for different value judgements (e.g., the choice of discount rate),
2. Set upper and lower bounds on the possible range of estimates. Depending upon the source of uncertainty or debate surrounding the estimations, this might be done by
   (a) considering empirical evidence from other research studies;
   (b) considering current practice in the literature;
   (c) soliciting judgements from those who will be making decisions based on the cost-effectiveness study.

3. Calculate study results based on combinations of the 'best guess', 'most conservative' and 'least conservative' estimates of the variables concerned.

The above procedure may result in a clear pattern emerging whereby one alternative dominates another in terms of cost-effectiveness, unless very pessimistic assumptions about its costs and effects are made. In other cases, especially those where there are wide ranges on a number of the estimates, no clear pattern may emerge. Here it might be necessary to specify a threshold set of estimates above or below which one alternative may no longer be more cost-effective than another. (The threshold estimates could then be presented to the decision-maker to assess the likelihood of this state-of-the-world occurring.) Alternatively, particularly in a situation where there is a clearly identified client for the study results, it might be possible to elicit some of his chosen values (such as the rate of discount) for use in the analysis. However, it should be pointed out that few evaluations in the health care field have reached a very high sophistication in this regard.

The level that most analysts should aim for is a clear identification of the uncertain or controversial estimates and a clear exposition of the ways in which different assumptions about these would impact on study results.

5.2. DESIGNING A COST-EFFECTIVENESS STUDY: EXERCISE

Imagine that you have been consulted on the following issue. Try to apply the knowledge you have gained so far.

5.2.1. Description of situation

Occasionally patients die from pulmonary embolism (i.e., clots in the blood vessels leading to the lung) following general surgery. Although death is relatively rare, the incidence is higher in patients over the age of 40 years who have undergone surgery lasting at least 30 minutes under general anaesthetic. Existing studies suggest that about 8 in 1000 patients will die.

The current approach is to treat postoperatively as and when venous thromboembolism becomes clinically apparent. The clinical signs might include (for pulmonary embolism) pleuritic chest pain, shortness of breath or coughing up blood, or (for deep-vein thrombosis, a related condition) pain and tenderness in the thigh or calf. Once the signs occur, the diagnosis is confirmed by lung scanning (for pulmonary embolism) or

Cost-effectiveness analysis

by venography (for deep-vein thrombosis). The treatment for both types of venous thromboembolism is the same—full-dose anticoagulant therapy consisting of heparin given intravenously for 7–10 days, followed by outpatient treatment with sodium warfarin for 12 weeks. Anticoagulant therapy prolongs hospital stay following surgery and some patients will have major bleeding complications.

Recently there has been an interest in prophylaxis. The options include the following.

1. *Primary prophylaxis*

(a) *Subcutaneous administration of heparin in low doses*. In this approach all patients would be given heparin subcutaneously for 2 hours preoperatively and then every 8 hours for 7 days postoperatively. If clinically suspected deep-vein thrombosis or pulmonary embolism develops, venography or lung scanning would be performed. If this confirms the diagnosis, full-dose anticoagulant therapy (with heparin) would be given. The low dose prophylaxis is not associated with significant clinical bleeding.

(b) *Intravenous administration of dextran*. In this approach all patients would be given dextran intravenously for 4 days postoperatively. If clinically suspected deep-vein thrombosis or pulmonary embolism develops, venography or lung scanning would be performed. If this confirms the diagnosis, full-dose anticoagulant therapy would be given, as in (a). Dextran prophylaxis carries slight risks of complications but these can be reduced to less than two per cent by careful administration.

(c) *Intermittent pneumatic compression of legs*. In this approach, an inflatable cuff is strapped to the patient's leg, enabling gentle pressure to be applied to the calf in a regular cycle. The procedure typically begins during the operation and is continued until the patient is considered no longer at risk, e.g., when the patient is ambulant. The cuff is worn continuously but is removed once per nursing shift. The modern devices that provide intermittent pneumatic compression are free of clinically significant side effects; in particular, there is no risk of bleeding.

As in (b), if clinically suspected deep-vein thrombosis or pulmonary embolism develops, venography or lung scanning would be performed. If this confirms the diagnosis, full-dose anticoagulant therapy (with heparin) would be given.

2. *Secondary prevention*

The approach here would be to perform leg scanning using iodine-125-labelled fibrinogen daily for 3 days following surgery and then on alternate days for up to 5 days, or up to the time of discharge if the patient is not ambulant. Leg scanning is free of complications.

If a positive scan is obtained, venography would be performed to confirm the diagnosis. Also lung scanning would be performed on patients showing clinical signs of pulmonary embolism. If the diagnosis is confirmed, patients would undergo full-dose anticoagulant therapy, as in the case of primary prophylaxis above.

5.2.2. Tasks

(A) Set out the five alternatives in a clear form, showing the sequence of diagnostic and therapeutic actions arising under each. (You may find that a diagrammatic representation helps.)

(B) Consider the following methodological issues, important in designing the cost-effectiveness study:
1. What should be the viewpoint for the study? (In particular, whose costs should be considered?)
2. What broad categories of cost should be considered for each alternative?
3. What would you choose as the main measure of effectiveness of the alternatives?
4. Are there any other attributes of the alternatives that should also be considered in addition to the main effectiveness measure?
5. What kind of medical evidence will be required for the cost-effectiveness study?
6. What are likely to be the main uncertain factors for which a sensitivity analysis might be required?

(Do not turn to Task (C) until Task (B) has been completed.)

Cost-effectiveness analysis

(C) Assume that the following data have been made available. Use them to calculate the cost effectiveness of the alternatives. Also, indicate the major points you would make in a discussion of the cost-effectiveness results.

Costs ($ 1982)

1. Prophylactic procedures (per patient)

Intermittent pneumatic compression of the legs	33
Leg scanning with iodine-125-labelled fibrinogen	85
Intravenous administration of dextran	103
Subcutaneous administration of heparin in low doses	20

2. Diagnostic procedures (per patient)

Venography	88
Lung scanning	117

3. Full-dose anticoagulant therapy (per patient)

Hospitalization costs (for 7 extra days at $290 per day)	2030
Intravenous heparin therapy	30
Laboratory tests	104
Warfarin therapy	10
Physician fees	35
Total	2209

Outcomes (obtained from controlled clinical trials and given for a cohort of 1000 patients receiving each regimen)

1. Current (no programme) approach

No. of patients with clinically suspected deep-vein thrombosis	40
No. of patients with clinically suspected pulmonary embolism	30
No. of positive venograms	19
No. of positive lung scans	14
Deaths	8

2. Subcutaneous administration of Heparin

No. of patients with suspected deep-vein thrombosis	10
No. of patients with suspected pulmonary embolism	10
No. of positive venograms	4
No. of positive lung scans	4
Deaths	1

3. Intravenous administration of dextran

No. of patients with suspected deep-vein thrombosis	20
No. of patients with suspected pulmonary embolism	10
No. of positive venograms	9

No. of positive lung scans	4
Deaths	1

4. Intermittent pneumatic compression of legs
   Outcomes as for subcutaneous administration of heparin, except that number of deaths not known. However, it is known that the approach is effective in preventing deep-vein thrombosis.

5. Leg scanning

No. of positive scans	135
No. of patients with suspected pulmonary embolism	15
No. of positive venograms	107
No. of positive lung scans	7

   Deaths (not known, although it is thought that leg scanning is effective in preventing fatal pulmonary embolism).

5.2.3. Solutions

1. *Description of the alternatives*

An algorithm of the clinical alternatives is given in Fig. 5.1. This diagrammatic representation is useful in getting a 'feel' for those clinical options involving not just one action, but a sequence of interrelated actions. In this example, the number of objective tests performed (i.e., venography and lung scanning) depends on the number of patients developing clinically suspected deep-vein thrombosis or pulmonary embolism under each approach. The number of patients receiving full-dose anticoagulant therapy will then depend on the results of the diagnostic tests.

The option involving leg scanning is somewhat different from the others in that it employs a further objective diagnostic test in order to identify possible cases at an early stage. It can be seen from the diagram that the cost of this strategy for 1000 patients will be crucially dependent on the number of positive leg scans and the number of these which are subsequently ruled out (as cases of DVT) by venography (the 'gold standard' test for DVT).

A slightly more advanced method of setting out complex sequences of clinical alternatives is the decision tree. This approach is gaining considerable popularity and is described in Fineberg (1980) and Weinstein and Fineberg (1980). A decision tree flows from left to right beginning with an initial clinical choice or decision (indicated by a box) on a defined category of patient (or cohort of patients). As a result of the decision made there will be outcomes of given prior probabilities, depicted in the decision tree at a chance node (indicated by a circle). The sum of prior probabilities at each chance node (e.g. p1 + p2 + p3) is equal to unity. Our example is redrawn in decision tree format in Fig. 5.2.

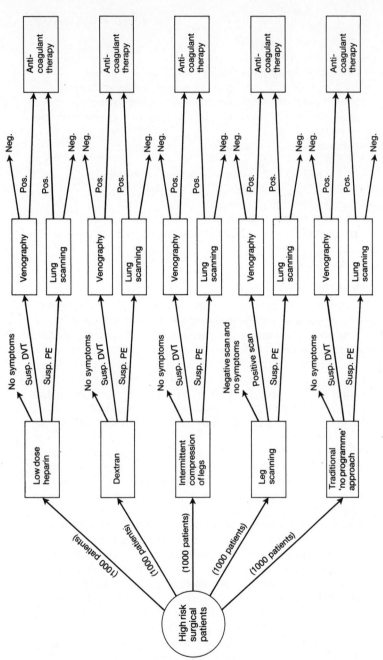

Fig. 5.1. Algorithm of clinical alternatives.

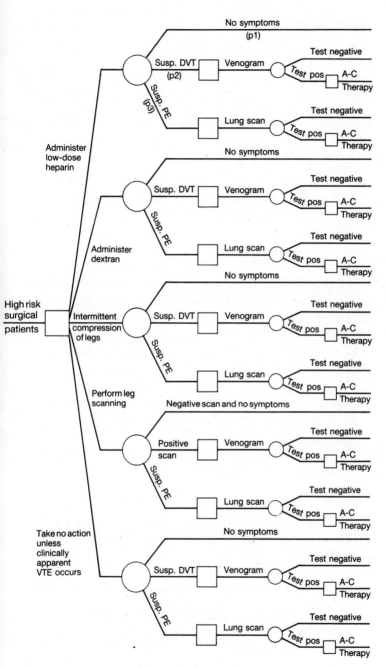

Fig. 5.2. Decision tree.

Cost-effectiveness analysis

2. *Methodological issues in study design*

(a) *Viewpoint for the study.* One viewpoint from which to undertake the analysis would be that of the third party payer. Therefore it would be most relevant to compare the hospital costs of the alternative regimens (including physician charges). Then the issue would be one of whether alternative (or broader) viewpoints would change the kind of results that merely a hospital cost comparison would give. For example:

From the patient's viewpoint
— is length of hospital stay the same?
— is recovery time the same?
— are health outcomes the same; death, complications?
— will expenditure be the same under all options, e.g. out-patient visits?

From the society's viewpoint
— are there any spillover costs or savings to other public or private sector agencies?

In this case, apart from the patient's interest in outcome there does not seem to be much conflict between the third party payer (Ministry of Health) and other viewpoints.

(b) *Categories of cost to be considered.* Obviously these depend on viewpoint, but one would definitely want to consider:

— hospital 'hotel' costs;
— prophylaxis costs (for the three preventive measures);
— treatment costs (i.e., full-dose anticoagulant therapy).

The main issues are:
— do we consider costs common to all the alternatives as well as the differences? (This affects mainly the costing of hospital in-patient stay.)
— how accurate are the hospital *per diem* costs?

(c) *Measure of effectiveness and other attributes of the regimens.* The main measure of effectiveness would obviously be deaths averted. (One might prefer life-years gained.) Other relevant attributes of the regimens include:

— unpleasantness of the diagnostic approaches, particularly venography;
— complications (e.g., bleeding) either from the prophylaxis or, more importantly, from the full-dose anticoagulant therapy;
— prolongation of hospital stay by anticoagulant therapy.

(d) *Source of medical evidence*. Ideally one would like evidence on the outcomes for each alternative, generated by controlled clinical trials. The variables that would be important to estimate include:

— the number of patients developing clinically suspected pulmonary embolism or deep-vein thrombosis;
— the number of positive venograms or lung scans (for those patients tested) and hence the number of patients receiving full-dose anti-coagulant therapy;
— the number of complications from therapy;
— the number of deaths from pulmonary embolism.

(e) *Factors requiring a sensitivity analysis*. This will depend on which of the medical parameters one can establish by randomized controlled trials—either in association with this study or drawn from other sources.

Obviously, cost-effectiveness of the regimens would be highly sensitive to the number of deaths resulting from the 'no programme' approach, the numbers averted by prophylaxis and the sensitivity and specificity of the leg scanning, venography and lung scanning procedures. Also, as primary prophylaxis involves giving everyone the low-dose therapy, the cost-effectiveness results would be highly sensitive to the cost of this. Finally, hospitalization costs are a large part of the total cost. Therefore it would be important to explore the sensitivity of results to variations in these costs and it would be worthwhile varying both the daily rate assumed and the prolongation of hospital stay assumed.

3. *Calculations of cost-effectiveness of the alternatives*

The data on the flow of patients under each regimen are added to the algorithm of clinical alternatives in Fig. 5.3. These data can then be combined with the cost data to give the total cost (per 1000 patients) for the five options. This cost is shown in Table 5.1, along with the effects, in terms of deaths. Table 5.1 also shows the incremental analysis in terms of costs and lives saved. It can be seen that the most cost-effective option is subcutaneous administration of heparin, which has the lowest costs yet saves seven lives.

Points that might be raised in a discussion of the results are

1. How sensitive is this result to the assumptions made about the cost of prophylaxis and the cost of hospitalization?
2. Is the effectiveness of intermittent pneumatic compression worth evaluating through a randomized controlled trial? (This approach is almost as cheap as subcutaneous administration of heparin, yet does not carry any risk of bleeding or wound complications due to prophylaxis.)

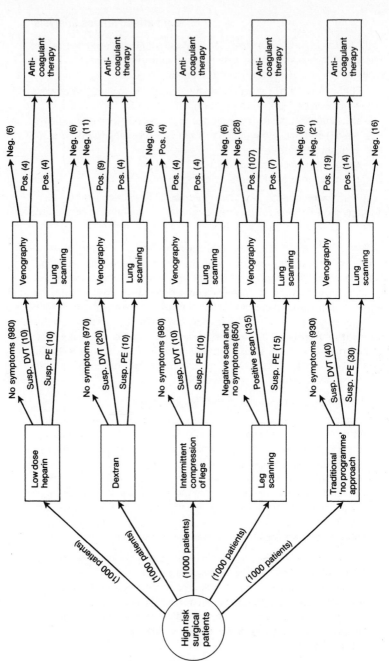

Fig. 5.3. Algorithm indicating the flows of patients under regimen.

Table 5.1. Alternative strategies for prevention of fatal pulmonary embolism in high-risk surgical patients

Strategy	Data		Incremental comparison		
	Cost/1000 ($)	Deaths/1000	Cost	Lives saved	C/E ratio cost/life saved
'No programme'	80 000	8			
Inraveneous dextran	135 000	1	55 000	7	7 900
Low-dose subcutaneous heparin	40 000	1	−40 000	7	<0
Intermittent pneumatic compression of legs	53 000	?	−27 000	?	?
Leg scanning with ^{125}I-fibrinogen	350 000	?	270 000	?	?

5.2.4. *A published study*

The situation upon which this exercise is based was, in fact, real. A group of Canadian researchers have designed and undertaken a cost-effectiveness study to address the issue concerned and this is published in the *Canadian Medical Association Journal* (Hull *et al*. 1982). The paper is assessed in Section 5.3 using the 10 questions set out in Annex 3.1. It is suggested that you locate the article and consider the comments below in the light of your own attempt at the problem and the solution given in Section 5.2.3.

5.3. CRITICAL APPRAISAL OF A PUBLISHED ARTICLE

Reference: Hull, R. D., Hirsh, J., Sackett, D. L., and Stoddart, G. L. (1982). Cost-effectiveness of primary and secondary prevention of fatal pulmonary embolism in high-risk surgical patients. *Can. Med. Assoc. J.* **127**, 990–5.

1. Was a well-defined question posed in answerable form?

__x__ YES _____ NO _____ CAN'T TELL

The authors considered both the dollar costs and the effects (deaths due to pulmonary embolism averted) of several strategies for preventing fatal pulmonary embolism in high-risk general surgical patients. The five alternatives compared are stated on p. 990, col. 2, of the article: (1) primary prevention with subcutaneous administration of heparin; (2) primary prevention with intravenous administration of dextran; (3) primary prevention with intermittent pneumatic compression of the legs; (4) secondary prevention with iodine 125-labelled fibrinogen leg scanning, and (5) treatment of clinically apparent thromboembolism. The viewpoint for the analysis could have been more explicitly stated. From p. 991, paragraph 2, col. 1, it appears that the analytical viewpoint is that of the third-party paying agency responsible for reimbursement of hospital and medical costs in the province of Ontario.

2. Was a comprehensive description of the competing alternatives given?

___x___ YES   _____ NO   _____ CAN'T TELL

The competing alternatives are reasonably well described. Details on the administration of heparin (p. 991, col. 1), dextran (p. 992, col. 1) and leg scanning (p. 992, col. 1) are provided in subsections of the article which describe the initial strategy and its subsequent investigations. (Unfortunately, the layout of the article headings might confuse a reader, since it appears that this information is being provided under a section entitled 'Costs of the strategies'.) Intermittent pneumatic compression of the legs (p. 992, col. 1) and treatment of clinically-apparent thromboembolism, called 'traditional (no programme) approach' (p. 992, col. 2) are dealt with similarly, though in less detail; however, the authors have provided references for all clinical protocols used. The list of competing alternatives appears complete and, as the authors have stated, option (5) is, in essence, the *do-nothing* alternative. In complex comparisons such as this, it is useful to visualize the comparison in terms of a flow chart of patients, perhaps even in a decision analytic form. It might have been helpful for readers if the authors had done so in this case. An example of such a chart, showing the clinical options and the distribution of patients among the various pathways was presented earlier in Fig. 5.3 above. The chart is based on the authors' use of an illustrative cohort of 1000 patients in each alternative.

3. Was there evidence that the programmes' effectiveness had been established?

___x___ YES   _____ NO   _____ CAN'T TELL

The authors have addressed the clinical evidence directly (p. 992, col. 2). Most of the evidence is drawn from well-referenced randomized controlled trials, some from the authors' own setting. It is noted that evidence on the effectiveness of leg scanning and intermittent compression was inferred from knowledge of venous thrombosis and its relationship to fatal pulmonary embolism, since no RCTs with fatal pulmonary embolism as an endpoint exist for these two alternatives. This lack of evidence is taken into account in subsequent analyses. However, it would make the analyses problematic only if either leg scanning or intermittent compression were likely to be *more effective* prophylactic strategies than heparin.

4. Were all important and relevant costs and consequences for each alternative identified?

___x___ YES   _____ NO   _____ CAN'T TELL

This may be a debatable assessment because the identification of consequences appears to be handled more clearly than that of costs. Deaths due to pulmonary embolism averted are clearly indicated as the effect of interest. Clinical complications of the alternative strategies also are identified, especially the risk of bleeding with heparin and the risk of anaphylactoid reaction and fluid overload with dextran (p. 993, col. 2). While detailed discussion of potential clinical complications is not provided, the authors have referenced their view that more explicit consideration of such complications would not change the basic results of the analysis.

With respect to the range of costs identified, the cost of each strategy is defined as 'the direct cost of the prophylactic procedure plus the diagnostic and treatment costs of nonfatal venous thromboembolism' (p. 991, col. 1). Although further detail is provided in Tables I and II of the article, the description of individual cost components could perhaps be clearer and more comprehensive. For example, it is not possible to ascertain whether capital costs are considered.

Other categories of costs and consequences, such as out-of-pocket costs, indirect costs and indirect benefits to patients, are excluded from the analysis because it has not been performed from a societal viewpoint. However,these would only be of significance if they were higher for primary prophylaxis (especially heparin) than for the traditional approach of waiting to treat clinically-apparent venous thrombo-embolism.

5. Were costs and consequences measured accurately in appropriate physical units?

___x___ YES   _____ NO   _____ CAN'T TELL

The measurement of deaths due to pulmonary embolism averted is straightforward. With respect to costs, the ideal presentation would give both the quantities of all resources used and the unit costs of each resource, prior to multiplying the two and summing across all resources or cost items in order to derive the total cost for any alternative. The authors have attempted to summarize the quantities of resources used up in textual description (p. 991–2) and in Tables I and II. While this could perhaps be more thorough, it may be unreasonable to expect journal editors to be interested in a more detailed presentation! The authors have dealt with the issue of shared costs (especially overheads) by separating items used differentially by patients with venous thromboembolism from other hospital cost items, and by measuring the former separately while implicitly accepting average *per diem* measurements (and values) for the latter. While more sophisticated methods exist for handling this problem, there does not appear to be a compelling case for their use in this instance.

6. Were costs and consequences valued credibly?

_____ YES _____ NO ___x___ CAN'T TELL

Since the analysis deals (appropriately for the specific clinical focus) with effects measured in natural units, the progression to valuation of these effects in terms of their dollar benefit or utility is not applicable. The reporting of the valuation of costs is handled less adequately than readers might expect, however. The only statement which directly deals with the issue suggests that costs 'are derived from the third party and operating costs incurred in a university teaching hospital in Ontario' (p. 991, col. 1). This leaves readers to make at least two assumptions, both of which may be warranted, but which should have been made explicit. They are that (1) the unit costs of specific items were based upon market values as represented by entries on hospital budgets, reimbursement schedules for specific procedures, or prevailing market prices for the prophylactic agents, and (2) no significant imputations or adjustments to these values were required for any reason. While possible variations in these values are handled partially in the following sensitivity analysis, more explicit reporting of the valuation procedures would seem appropriate.

7. Were costs and consequences adjusted for differential timing?

_____ YES ___x___ NO _____ CAN'T TELL

Costs and consequences are not discounted to present values. However, discounting is inappropriate in the case of this study, since all costs and effects relevant to the analysis, as framed by the comparison statement and viewpoint, occur in the 'present'. That is, the analysis is conducted at one point in time, and the analytic horizon, from the beginning of the interventions to their resolution in outcomes of interest, is well inside one year.

Cost-effectiveness analysis

8. Was an incremental analysis of costs and consequences of alternatives performed?

___x___ YES _____ NO _____ CAN'T TELL

The presentation of the results provided at the bottom of p. 992 and top of p. 993 does this implicitly. However, the explicit presentation of the incremental analysis could be significantly improved. The increment in effectiveness associated with primary and secondary prophylaxis strategies is the number of deaths due to pulmonary embolism averted. This is found in the text. The increment in cost associated with primary and secondary prophylaxis is the difference in the total cost per 1000 patients between each alternative and the traditional (no-programme) approach. Incremental cost is rather tersely reported in the text at the top of p. 993: 'The traditional approach costs twice as much as subcutaneous heparin prophylaxis and about half as much as intravenous dextran prophylaxis.' The incremental analysis probably warrants a separate table, space permitting. An example of such a table was given earlier (Table 5.1). The use of a hypothetical cohort of 1000 patients managed by each of the clinical strategies facilitates considerably the presentation of the results.

9. Was a sensitivity analysis performed?

___x___ YES _____ NO _____ CAN'T TELL

Sensitivity analysis is performed on several variables, as reported in Tables III and IV, p. 993. No specific justification is provided for the ranges of variables employed. The quite wide variation in cost values does, however, seem to deflect some of the criticism made above, since the cost-effectiveness of heparin is a relatively robust result. Of particular note is the joint possibility that the costs of prophylaxis have been underestimated and hospitalization costs overestimated in the initial analysis, since this would bias the analysis against the traditional (no programme) approach. Sensitivity analysis on these assumptions simultaneously (Table IV, last column), rather than one-at-a-time as is typically done, showed that heparin became only slightly more costly than the traditional approach, while still saving seven lives.

10. Did the presentation and discussion of study results include all issues of concern to users?

__x__ YES _____ NO _____ CAN'T TELL

The analysis does not explicitly provide cost-effectiveness ratios for the alternatives; rather, it discusses directly the large incremental effectiveness of heparin and the likely cost saving which would accompany its use. The results are not compared to those of other investigators since this is the initial cost-effectiveness analysis of these methods. Based on the limited task of this analysis and the nature of the recommended strategy of subcutaneous heparin prophylaxis, further issues of generalizability, ethics, distributional considerations such as equity, and implementation would not appear problematic and are, therefore, not addressed in any detail by the authors.

5.4. INCREMENTAL ANALYSIS OF COSTS AND EFFECTS: EXAMPLE

It is possible to make a number of errors in comparing cost-effectiveness ratios between programmes. The following examples have been constructed to illustrate some common errors. Although none of these errors were made by Logan, Milne, Achber, Campbell, and Haynes (1981), we have taken the data from their study and modified them to create the following examples. (Note: this section is technical and can be omitted without loss of continuity.) The main points are that it is important: (i) to be clear on the alternative programmes being compared, including the implicit 'do nothng' option; (ii) to consider whether 'doing nothing' will itself have costs and effects, as the size of these may alter the relative cost-effectiveness of other options; (iii) to consider the effect that a budgetary limit may have on option selection.

5.4.1. Single programme, single level

A = Hypertension screening at work and treatment at work by a nurse practitioner (the worksite (WS) programme in Logan's paper).

E (Effect) = average first year diastolic BP reduction in mm Hg.

C (Cost) = average cost to find plus treat for one year.

Cost-effectiveness analysis

(a) Tabular display   Graphical display

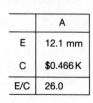

	A
E	12.1 mm
C	$0.466 K
E/C	26.0

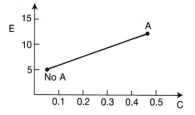

Assumptions
1. E and C are defined and measured properly.
2. Without programme A there would be no cost and no effects, i.e., no hypertensives would be found and treated, and the untreated hypertensives would have no change in their diastolic BP over the year

Decision
Is 26.0 mm Hg reduction/$1K worthwhile?

(b) Display if Assumption 2 is violated. Note that the E/C ratio is reduced.

	A	No A	A-No A
E	12.1	5.0	7.1
C	0.466	0.050	0.416
E/C	–	–	17.1

Decision
Is 17.1 mm Hg reduction/$1K worthwhile?

5.4.2. Two programmes, single level each

A = Hypertension screening at work and treatment at work by a nurse practitioner (WS programme).

B = Hypertension screening at work and treatment by community general practitioner [the regular care (RC) programme in the paper by Logan *et al.*]

Cost-effectiveness analysis

(a) Tabular display Graphical display

	A	B
E	12.1	6.5
C	0.466	0.434
E/C	26.0	15.0

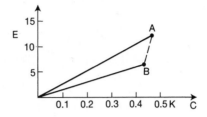

Assumptions
1. E and C are defined and measured properly.
2. Without A or B there would be *no* costs and *no* effects.

Result
    A is more cost-effective than B. That is, for a given amount of cost, A will produce greater average BP reductions. (Note: for the 'scale' of programme tested.)

(b) Display if Assumption 2 is violated (X = no formal programme)

	A	B	X	A–X	B–X	A–B
E	12.1	6.5	2.0	10.1	4.5	5.6
C	0.466	0.434	0.410	0.056	0.024	0.032
E/C	–	–	–	180.4	187.5	175.0

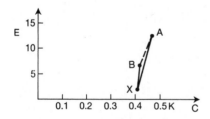

Note, this can reverse the result. B is now more cost-effective than A. (Note, however, A may still be chosen if 175 mm Hg reduction per $1K is judged worthwhile and if there is sufficient budget.)

(c) Logan *et al.* (1981) version. X, the no programme, is assumed to have E = 5, C = 0.

Cost-effectiveness analysis

	A	B	X	A–X	B–X	A–B
E	12.1	6.5	5.0	7.1	1.5	5.6
C	0.466	0.434	0.0	0.466	0.434	0.032
E/C	–	–	–	15.2	3.5	175.0

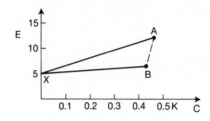

Result

A is more cost-effective than B.

5.4.3. Two programmes, multiple levels of each

By treating more aggressively, better compliance and better results may be obtained. Let A1, A2, A3 represent increasingly aggressive versions of programme A. Similarly for B. Note, this is a new example. The data do not match those in the previous discussion.

Per patient data:

	A1	A2	A3
E	12.2	14.0	16.0
C	0.680	0.800	1.0
E/C	17.9	17.5	16.0

	B1	B2	B3
E	6.5	9.1	23.0
C	0.540	0.740	1.500
E/C	12.0	12.3	15.3

Marginal (incremental) data—to evaluate the extra aggressiveness:

	A1	A2–A1	A3–A2
E	12.2	1.8	2.0
C	0.680	0.120	0.200
E/C	17.9	15.0	10.0

	B1	B2–B1	B3–B2
E	6.5	2.6	13.9
C	0.540	0.200	0.760
E/C	12.0	13.0	18.3

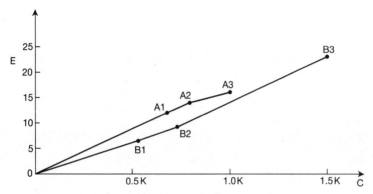

Results

1. A1 is the most cost-effective alternative.
2. However, it may *not* be worthwhile to do A1, or it may be worthwhile to do *more* than A1.

For example: consider a community in which there are 1000 hypertensives. Scale up the data to community figures by multiplying E and C by 1000. (Note that assumptions are required here regarding programme returns to scale.)

(a) Assume it is feasible (technically and politically) to screen only part of the community, if you wish, and to treat some patients with one method and some with another. The optimal plan lies somewhere on the line joining 0, A1, A2, B3 (the efficient frontier), depending on the dollar limit (L) available:

Community cost	Optimal programme mix	Marginal E/C
L ≤ 680K	A1 for some; 0 for remainder	17.9 (A1–0)
L > 680K ≤ 800K	A2 for some; A1 for remainder	15.0 (A2–A1)
800K < L ≤ 1500K	B3 for some; A2 for remainder	12.9 (B3–A2)

103

Cost-effectiveness analysis

(b) Assume you must screen and treat all or none, and those treated must be treated the same way, i.e., your only choices are points 0, A1, A2, A3, B1, B2, B3, on a community-wide basis.

(1) *Given a standard if*	*Choose*	(2) *Given a dollar limit*(L) *if*	*Choose*
17.9 < Std	0	L < 540K	0
15.0 < Std ⩽ 17.9	A1	540K ⩽ L < 680K	B1
12.9 < Std ⩽ 15.0	A2	680K ⩽ L < 800K	A1
Std ⩽ 12.9	B3	800K ⩽ L < 1000K	A2
		1000K ⩽ L < 1500K	A3
		1500K ⩽ L	B3

Note (a) that B2 is dominated and is never on any list; (b) the difference between lists (1) and (2); and (c) the example is even more complicated if there are differences due to programme scale.

5.5. CRITICAL APPRAISAL OF A PUBLISHED ARTICLE

Reference: Logan, A. G., Milne, B. J., Achber, C., Campbell, W. P., and Haynes, R. B. (1981). Cost-effectiveness of a worksite hypertension treatment program. *Hypertension* **3** (2), 211–18.

This paper is assessed below using the 10 questions set out in Annex 3.1. It is suggested that you locate the article and attempt the exercise before reading the assessment.

1. Was a well-defined question posed in answerable form?

__x__ YES _____ NO _____ CAN'T TELL

The study was undertaken to compare the cost-effectiveness of a worksite-based hypertension programme in which all care was provided onsite, with the normal practice of management of hypertension by community physicians.

The analysis is conducted from the combined viewpoint of the health care system and of the patients involved in the study. This combined viewpoint should not be confused with a 'societal' viewpoint, however, since the latter would necessitate consideration of at least the employers' viewpoint as well as that of government and patients.

2. Was a comprehensive description of the competing alternatives given?

___x___ YES _____ NO _____ CAN'T TELL

A detailed description of the worksite care (WS) programme is provided on p. 212, col. 1 of the paper, with reference to the standard protocol used by the nurses to manage these patients. Less detail was provided for the alternative, regular care by physicians in private practice, although it is likely that a clinical audience reading this article would be familiar with standard treatment practices for hypertension. Moreover, a generalized description of the management of patients in the community setting would be difficult to provide since the specific approach to treatment depends on both the individual physician and patient.

Although a *do-nothing* alternative is never explicitly considered as a viable option, the authors do examine the effects (but not costs) of providing no treatment in their later calculations.

3. Was there evidence that the programmes' effectiveness had been established?

___x___ YES _____ NO _____ CAN'T TELL

The economic evaluation was conducted concurrently with a randomized trial of the two treatment alternatives. Standard protocol for an RCT was observed (i.e., prognostic stratification, randomization, description of inclusion criteria, blinding of assessor, reporting of loss to follow-up, etc.). According to the criterion of statistical significance, both alternatives were effective in significantly reducing diastolic blood pressure (BP). However, the WS programme was relatively more effective (12.1 ± 0.6 mm Hg reduction in BP) than the RC programme (6.5 ± 0.6 mm Hg reduction in BP).

It is not clear, though, that the observed reductions in BP were due entirely to either of the treatment programmes. The authors mention in the *Discussion* section that some of the observed reduction in BP might be the result of patient familiarity with the measurement procedure and regression toward the mean of the BP in the general population. Citing previous findings, they suggest that diastolic BP might conceivably fall by 5 mm Hg in one year in the absence of any treatment. They do not comment, however, on whether the remaining effect beyond the estimated natural reduction of 5 mm Hg (7.1 mm Hg for the WS group, and 1.5 mm Hg for the RC group) is statistically significant.

4. Were all the important and relevant costs and consequences for each alternative identified?

___x___ YES   _____ NO   _____ CAN'T TELL

The majority of costs and consequences resulting from the two alternatives have been identified by the authors. The costs fall into two categories: costs attributed to screening and costs attributed to therapy. Both of these categories include costs which were borne either by the health care system or by patients themselves.

Screening costs included personnel, equipment and supplies, travel, participants' time, and administrative costs, while treatment costs were composed of the provision of care and laboratory examinations, hospitalization and drugs (see pp. 212–13).

The viewpoint of the employer is missing from the analysis, as indicated by the omission of any WS-related costs (such as costs of the WS room, electricity, furnishings, etc.). The authors later claim that '... health care facilities are already available in most places of work with 200 or more employees ...', thereby implying that no costs are incurred with the set-up of the WS facility. But to assign a zero cost to the facility implies that the opportunity cost of using it is nil, or in other words, that nothing will be sacrificed (i.e., no competing programmes, treatment, educational endeavours, etc.) by devoting the space to screening and treating hypertension.

It is important to note that costs associated with the conduct of the trial and evaluation (as opposed to the costs of screening and treatment) have been omitted. These costs were legitimately excluded because they would not constitute ongoing (or 'recurrent') costs of either of the programmes once implemented.

The effect identified in the analysis was the average reduction in diastolic BP. Although this is undoubtedly the common outcome of interest for the two programmes, the WS programme resulted in higher drug use for its hypertensive patients. No mention is made of the possibility that greater drug use in the WS cohort resulted in more frequent or more severe side effects.

5. Were costs and consequences measured accurately in appropriate physical units?

___x___ YES   _____ NO   _____ CAN'T TELL

The measurement of costs in particular and of effects was quite detailed, and whenever possible done on an individual patient basis. Personnel costs for screening were obtained from weekly logs which monitored travel, service and office time. Equipment costs were itemized and an annualized equipment value was calculated for each item. The costs of the participants' time included imputations for waiting, service and questionnaire completion time at screening visits. Administrative costs were estimated to be 30 per cent of the health care system cost of the screening programme. Presumably this percentage was based on some prior knowledge about the proportions of total cost attributable to administration; however there is no mention of how this percentage was derived.

Treatment costs for the provision of services by physicians (i.e., diagnostic and therapeutic procedures) for diagnostic radiology and laboratory tests, and for the number of days of hospitalization were obtained from service files kept by the Ontario Health Insurance Plan (OHIP). The files were reviewed so that only services properly identified as related to hypertension were included. Nurses' time was measured using weekly logs of their total time spent in patient care activities (including direct care, travel and paperwork related to patient care). Drug cost data were obtained from insurance companies who sponsored drug insurance programmes in industries involved in the study.

Logs and encounter forms were used to measure patient inputs into the programmes. An average time per visit was calculated, based on distance travelled and time spent in travel, waiting, and service. This average was then multiplied by the number of visits for these services by the individual during the study year, as determined from their OHIP service file. In the final calculations, averaged values for individual cost items were substituted for missing data (although only complete data were used to compare results of individual cost items between groups).

One area of cost measurement which was unsatisfactory was the assessment of 'shared' resources. Items such as the WS facility and laboratory equipment were ignored. However, even if these resources were not devoted entirely to either programme, some portion of their cost ought to be attributed to the programmes concerned.

The effect of interest, the average reduction in diastolic BP, was measured by subtracting the endpoint BP level from the entry level. Entry BP was calculated by averaging all diastolic BP measurements at the first and second BP screens, while endpoint BP was established by averaging all diastolic BP measurements at the year-end assessment.

Cost-effectiveness analysis

6. Were costs and consequences valued credibly?

__x__ YES _____ NO _____ CAN'T TELL

Prevailing market prices were generally used to value the resource inputs to the programmes and the sources for these values were clearly identified. Personnel costs were valued using the appropriate salary scales; for nurses, prorated annual salaries were used (plus fringe benefits), while the Ontario Medical Association fee schedule was employed to calculate physicians' inputs. Actual wage values were used in estimating the cost to patients of time lost from work, and where these data were unavailable, average wage values were substituted.

Although the authors claim to measure the cost to the patient of lost leisure time (p. 217, col. 1), the use of wage rates to estimate this cost may not be an accurate assessment of the true opportunity cost of lost leisure. People may place a higher value on their leisure time in the early evening, thus the need to 'induce' workers to sacrifice leisure time by paying them a higher than normal wage to work extra hours. Patients' costs for the RC group may, therefore, be underestimated if much of the time required for treatment of these patients occurred after work.

Oddly, the authors come to exactly the opposite conclusion. They state that, 'Since loss of leisure time may not represent a cost to society (no effect on worker productivity), it may be argued that patient cost in the RC group should be valued at some fraction of the patient cost in the WS group' (p. 217). But lost leisure time certainly represents a cost (in terms of lost utility) to the patient. Failure to recognize this item as a cost from the patient's viewpoint may result in the implementation of a seemingly cost-effective programme in which no-one is willing to participate. This point illustrates the importance of recognizing that different costs are associated with different viewpoints; consequently a programme which appears attractive from one viewpoint may not necessarily meet with success in its implementation.

Per diem rates were used to value hospitalization costs. However, these average rates are by definition generally not good indicators of the true resource cost of treating a specific disease. But, as the authors point out, the time and expense involved in utilizing more sophisticated methods of hospital cost allocation is likely not warranted due to the small incidence of hospitalization.

7. Were costs and consequences adjusted for differential timing?

_____ YES   x   NO   _____ CAN'T TELL

No discounting of future costs and effects was employed, but it was not necessary due to the one-year duration of the study.

8. Was an incremental analysis of costs and consequences of alternatives performed?

x   YES   _____ NO   _____ CAN'T TELL

Table 4 presents the average cost-effectiveness ratios for the two programmes and the *incremental* cost-effectiveness of the WS programme over the RC programme. The incremental cost of lowering BP in the WS programme (i.e., the cost over and above RC) was $5.63 per mm Hg reduction. The authors then compare this ratio to the average for the RC and conclude that, 'If conventional treatment of hypertension (RC) is considered worthwhile, it is clearly more cost-effective to replace RC with WS treatment for the target group identified in this study' (p. 215). However, caution should be exerted when comparing incremental ratios to averages to determine the merit of the programmes. As the authors note, this practice presupposes that the RC programme has been demonstrated to be worthwhile. However, their study is designed to compare RC to WS programmes and not to compare the RC programme to a *do-nothing* alternative. Thus although this study demonstrates the relative cost-effectiveness of WS over RC, it says nothing about the value of RC *per se*.

The authors do consider the *effects* of a *do-nothing* alternative, however. Table 6 presents revised results assuming that a 5 mm Hg reduction in BP would have occurred with no treatment, and these results are plotted in Fig. 2. However, nowhere is there recognition of the possibility that non-zero costs may be associated with the do-nothing option. The existence of costs, as well as effects, arising from *no treatment* may lead to fundamentally different conclusions and ranking of priorities, as demonstrated in Section 5.4 above.

9. Was a sensitivity analysis performed?

x   YES   _____ NO   _____ CAN'T TELL

A sensitivity analysis was conducted to test the robustness of the results to variations in the assumptions used to calculate entries for missing data.

Cost-effectiveness analysis

The authors purposely chose values for the sensitivity analysis which would *overestimate* the cost of the WS programme and *underestimate* the cost of the RC programme. Thus, whenever data were missing for the WS group, the maximum cost encountered in that group was substituted. Similarly, whenever data were missing for the RC group, minimum values were substituted. The results of the sensitivity analysis indicated that, even when extreme assumptions were used which unrealistically favoured the RC programme, the incremental cost-effectiveness ratio was still less than the average cost-effectiveness ratio for RC.

10. Did the presentation and discussion of study results include all issues of concern to users?

  x   YES   NO   CAN'T TELL

The conclusion of this study is that treatment of employed hypertensives at their place of work is both more effective and more cost-effective than usual care in the community. The authors have also broken down their results to show that the cost implications of the programme differ according to the viewpoint undertaken. Although the costs to the health care system of the WS programme exceed the costs of RC, the costs to the patient for the WS programme are lower than those for RC. This type of subanalysis (of the burden of costs) is important because even though one programme may be more cost-effective overall than another, if the costs to the participants in the programme are prohibitive, obstacles to its implementation may arise.

The authors also discuss the attitudes of the participants to the WS programme; although initially wary of the WS programme (as evidenced by the need to consult a community physician at the outset), the WS group eventually accepted the medical care provided at the worksite.

The advantages of using the work setting to manage hypertension, aside from its greater cost-effectiveness, were also presented. The authors cite the ease of set-up due to existing facilities, the facilitation of access to care for a population for whom usual care in the community may be inconvenient, and the treatment of previously detected but untreated hypertension as some of the potential benefits of the WS programme.

Finally, although the issue of generalizability was never explicitly discussed, the authors clearly emphasize that their findings are applicable specifically to the target group identified in the study. The extent to which the results may vary, according to the worksite setting, the target group, the existence of similar programmes, or even the health care system, are not known.

REFERENCES

Baynham, R. A., Dent, P. B., and Torrance, G. W. (1984). A cost analysis of alternative methods of administering gammaglobulin for the treatment of antibody deficiency syndrome. QSEP Research Report No. 93, Faculty of Social Science, McMaster University, Hamilton, Ontario.

Bush, J. W. (1973). Discussion. *Health status indexes* (ed. R. L. Berg). Hospital Research and Educational Trust, Chicago.

Department of Epidemiology and Biostatistics (1981). Clinical Epidemiology Rounds: How to read a clinical journal. V: To distinguish useful from useless or even harmful therapy. *Can. Med. Assoc. J.* **124**, 1156–62.

Drummond, M. F. and Stoddart, G. L. (1984). Economic analysis and clinical trials. *Controlled Clinical Trials* **5**,115–128.

Evans, R. G. and Robinson, G. C. (1980). Surgical day care: measurements of the economic payoff. *Can. Med. Assoc. J.* **123**, 873–80.

Fineberg, H. V. (1980). Decision trees: construction, uses, and limits. *Bull. Cancer* **67**, 395–404.

Grossman, M. (1972). *The demand for health*. National Bureau of Economic Research, New York.

Hagard, S. and Carter, F. A. (1976). Preventing the birth of infants with Down's Syndrome: A cost benefit analysis. *Br. Med. J.* **i**, 753–6.

Henderson, J. B. (1982). An economic appraisal of the benefits of screening for open spina bifida. *Social Science and Medicine* **16**, 545–60.

Hull, R., Hirsh, J., Sacket D. L., and Stoddart, G. L. (1982). Cost-effectiveness of primary and secondary prevention of fatal pulmonary embolism in high-risk surgical patients. *Can. Med. Assoc. J.* **127**, 990–5.

Keeler, E. and Cretin, S. (1983). Discounting of life savings and other nonmonetary effects. *Management Science* **29**, (3), 300–6.

Logan, A. G., Milne, B. J., Achber, C., Campbell, W. P., and Haynes, R. B. (1981). Cost-effectiveness of a worksite hypertension treatment programme. *Hypertension* **3**, (2), 211–18.

Quade, E. S. (1982). *Analysis for public decisions* (2nd edn). North Holland, New York.

Weinstein, M. C. and Fineberg H. V., (1980). *Clinical decision analysis*. W. B. Saunders Company, Philadelphia.

—— and Stason, W. B. (1977). Foundations of cost effectiveness analysis for health and medical practices. *N. Engl. J. Med.* **296**, 716–21.

6. Cost–utility analysis

6.1. SOME BASICS

Cost–utility analysis is a form of economic appraisal that focuses particular attention on the quality of the health outcome caused or averted by health programmes or treatments. It has many similarities to cost-effectiveness analysis, and thus all the points discussed in Chapter 4 on cost analysis and many of those discussed in Chapter 5 on cost-effectiveness analysis also apply here. The first section of this chapter reviews some of the general issues the evaluator would need to consider when undertaking a cost–utility analysis. Later sections discuss particular issues in more detail.

6.1.1. How does cost–utility analysis differ from cost-effectiveness analysis?

In cost-effectiveness analysis (CEA) the incremental cost of a programme, from a particular viewpoint, is compared to the incremental health effects of the programme, where the health effects are measured in natural units related to the objective of the programme, e.g. cases found, cases of disease averted, lives saved, life-years gained. The results are usually expressed as a cost per unit of effect. In cost–utility analysis (CUA) the incremental cost of a programme, from a particular viewpoint, is compared to the incremental health improvement attributable to the programme, where the health improvement is measured in quality-adjusted life-years (QALYs) gained. The results are usually expressed as a cost per QALY gained. Thus, there are many similarities between CEA and CUA. For example, the questions of whether or not to include indirect costs and benefits (Section 5.1.3) and whether or not to include the costs of other treatments or programmes (Section 5.1.4) still apply.

The differences between the two techniques exist in the measurement of health outcomes. Both techniques require valid effectiveness data (from the literature, from your own study or from expert judgement supplemented by sensitivity analysis), but in the case of CUA only final output effectiveness data will suffice (e.g., lives saved, disability-days

112

averted). Intermediate output data (e.g., cases found, patients appropriately treated) are unsuitable, since they cannot be converted into QALYs gained. As an aside, intermediate outcomes may well be suitable for clinical decision analysis using a patient's utilities; they are simply unsuitable for CUA where the outcomes must be expressed in QALYs gained. By converting the effectiveness data to a common unit of measure, QALYs gained, CUA is able to incorporate simultaneously both the increase in the quantity of life (reduced mortality) and the increase in the quality of life (reduced morbidity). The quality adjustment is based on a set of values or weights called utilities, one for each possible health state, that reflect the relative desirability of the health state.

Because of the similarities between CUA and CEA some authors do not distinguish between the two. For example, Weinstein and Stason (1977) treat cost–utility analysis as a particular case of cost-effectiveness analysis. Thus, be aware in reading the literature that CUA may appear under other labels.

6.1.2. When should CUA be used?

The following are a number of situations where you might wish to use CUA:

1. When quality of life is *the* important outcome. For example, in comparing alternative programmes for the treatment of arthritis, no programme is expected to have any impact on mortality, and the interest is focussed on how well the different programmes improve the patient's physical function, social function and psychological well-being.
2. When quality of life is *an* important outcome. For example, in evaluating neonatal intensive care for very-low-birth-weight infants not only is survival an important outcome, but also the quality of that survival is critical.
3. When the programme affects both morbidity and mortality and you wish to have a common unit of outcome that combines both effects. For example, estrogen therapy for menopausal symptoms improves the quality of life by eliminating the discomfort of the symptoms, reduces mortality from hip-fractures, but increases mortality from complications such as endometrial cancer, uterine bleeding and endometrial hyperplasia, and gall bladder disease (Weinstein 1981).
4. When the programmes being compared have a wide range of different kinds of outcomes, and you wish to have a common unit of

113

output for comparison. For example, if you are a health planner who must compare several disparate programmes applying for funding, such as an expansion of neonatal intensive care, a programme to locate and treat hypertensives, and a programme of antepartum prevention of Rh-immunization of pregnant women.

5. When you wish to compare a programme to others that have already been evaluated using cost–utility analysis.

6.1.3. When should it not be used?

The following are situations when CUA should not be used:

1. When only intermediate output effectiveness data can be obtained. For example, in a study to screen employees for hypertension and treat them for one year, Logan, Milne, Achber, Campbell, and Haynes (1981) used end points of mm Hg blood pressure reduction. Intermediate outcomes of this type cannot be converted into QALYs for use in CUA.

2. When the effectiveness data shows that the alternatives are equally effective. In this case cost-minimization analysis is sufficient; cost–utility analysis is not needed.

3. When the quality of life is important but it can be captured by a single variable measured in easily understood natural units. For example, several alternative new methods for treating leg fractures might be compared in terms of the reduction of restricted activity days.

4. When the extra cost of obtaining and using utility values is judged to be in itself not cost effective. For example, this might be the case if cost-effectiveness analysis showed one alternative to be overwhelmingly superior, so much so that the incorporation of utility values at some considerable effort could almost certainly not change the result. Alternatively, CEA might show one alternative to be only somewhat superior, but it is known that the direction of the quality improvement also favours the superior alternative. In this case the extra effort to peform a CUA would simply reinforce the conclusion of the CEA, and therefore might very well not be undertaken. Of course, one could argue that essentially you must have performed a rough cost–utility analysis using judgemental utilities in order to reach these conclusions.

6.1.4. How are the utility values determined?

There are three methods of obtaining utility values for health states in cost–utility analysis: judgement, values from the literature or values from measurements on a sample of subjects.

Probably the simplest method is to use judgement to estimate the utility values, or a range of plausible values, and then to undertake extensive sensitivity analysis to explore the implications of this judgement. Judgements can be simple estimates made by the analyst or by a few physicians (for example, see Weinstein 1981), or they can be formal measurements made on a convenience sample of physicians or other experts [for example, see Torrance, Sackett, and Thomas (1973), or Pliskin, Shepard, and Weinstein (1980)].

In some cases it may be possible to use existing utility values available in the literature. Of course, it would be important to determine that the health states used in the measurement match those of your own study, that the subjects used in the measurement are appropriate to your own study and that the measurement instruments used were credible. Some utilities available from the literature are presented in Section 6.2.

Generally, the most accurate way to obtain utility values for your own study will be to measure them yourself. There are three instruments in common use: rating scale, standard gamble and time trade-off. Each of these is described in some detail in Section 6.3.

6.1.5. Whose utility values are appropriate?

There are two basic approaches to measuring the utility of health states. One method is to find people with the health state and measure their utility for their condition [for examples of measurements made on kidney dialysis patients see Sackett and Torrance (1978), and Churchill, Morgan, and Torrance (1984b)]. The other approach is to describe the condition, usually by an abbreviated written scenario, to people who do not have the condition and measure their utility for it. (For examples of this approach see Sackett and Torrance 1978, and Torrance, Boyle, and Horwood 1982.) A variation of this latter approach is to use subjects already knowledgeable of the health states concerned, for example physicians or nurses, to minimize the need for elaborate scenarios.

But who *should* you ask? The answer can be determined, in part, from the viewpoint of the study. Most cost–utility analyses are conducted from the societal viewpoint and are pertinent to public policy decisions. In this case, the appropriate utilities are those of an informed member of the general public or community representative. Informed means that the subject truly understands what the health state is like. This is the sticking point. A short point-form description of the health state, as used by many investigators (Kaplan, Bush, and Berry 1976), may not be sufficient for some subjects. However, long comprehensive descriptions, which have also been used (Sackett and Torrance 1978), may have greater face validity but may simply overload the cognitive abilities of the subject so that he or she merely latches on to a few key phrases and ignores the rest. Some investigators have used audio tapes (McNeil, Weichselbaum, and Pauker 1981; Boyd, Sutherland, Ciampi, Tibshirani, Till, and Harwood 1982) and even video tapes (Cadman and Goldsmith 1982) to supplement the descriptions. The issue of whether the level of detail in the presentation matters is still unresolved. Studies have shown both that it does (Boyd *et al*. 1982) and that it does not (Cadman and Goldsmith 1982). Finally, it is well known (Kahneman and Tversky 1982; Hershey, Kunreuther, and Shoemaker 1982) that the way in which a health state is described and the way in which the question is framed can systematically bias the answer. Given the uncertainties discussed above, the best current advice in measuring utilities on the general public is probably to use abbreviated descriptions to avoid cognitive overload, to supplement these with prior more detailed descriptions of the key phrases used in the abbreviated scenarios, and to avoid the framing bias by wording the question in a balanced (positive and negative) manner.

Because of the difficulties in accurately describing some health states to the general public, some researchers have used patients as the source of utilities for CUA (Churchill, Lemon, and Torrance 1984a). This has the advantage of eliminating the problem of health state description, but it may introduce other biases and problems. For example, patients are in a potential conflict of interest situation. Collectively, patients of a particular disorder have an incentive to exaggerate the disutility of their condition in order to enhance the cost–utility of preventive and treatment programmes directed at the disorder. Moreover, if it is desired to ultimately compare the cost–utility results of different programmes competing for the same scarce resources, it may be important to have the utility values determined on one set of subjects or at least on similar subjects rather than disparate groups of patients (for example, angina patients and scoliosis patients may be quite different in their age–sex profiles).

Finally, the use of patients as the source of utilities restricts the measurement to one health state utility per subject—the utility of his or her health state. Measurements on the general public, on the other hand, normally acquire many health state utilities during the one interview. Despite the disadvantages discussed above, patients are an appropriate source of preferences where the alternatives being compared are all directed at the same disorder, where utilities for this disorder are the only utilities required and where the primary intention is not to compare the results to programmes directed at other disorders.

Health professionals, such as physicians and nurses, have also been used as the source of health state utilities. This has many of the same advantages and disadvantages as the use of patients. It minimizes the problems of describing the states, but at the expense of possible bias due to conflicts of interest and due to the special age, sex and socioeconomic status of health professionals.

Who you should ask is only an issue if different groups are known to give different results. This has generally not been the case. Most investigators have found no difference among different groups—age, sex, general public, physicians, nurses, patients (Kaplan and Bush 1982; Sackett and Torrance 1978; Wolfson, Sinclair, Bombardier, and McGeer 1982), a few have small differences (Sackett and Torrance 1978) and none have found large differences.

6.1.6. How does one assess the validity of utility values?

The utility values in your study are valid if:

(a) the subjects are appropriate;
(b) the health state descriptions are adequate to properly describe the states and are neutral in their influence on the measurement;
(c) the measurement questions are framed in a balanced or neutral way; and
(d) the measurement technique itself is reliable and valid.

The first three of these points were discussed in Section 6.1.5. The fourth issue, of reliability and validity of the measurement techniques, is discussed here.

A measurement technique is reliable if it is consistent—if the same phenomenon can be measured a second time with identical results. Internal reliability refers to a second measurement taken as part of the original interview. The coefficient of internal reliability (as measured by the product moment correlation coefficient, r) for the rating scale ranges from 0.86 to 0.94 (Torrance *et al.* 1982), and for the standard gamble 0.77 to 0.92 (Torrance 1976; Torrance *et al.* 1982). These would all be judged acceptable. Test–retest reliability refers to a second measurement taken sometime later. A six-week test–retest r in the range of 0.63 to 0.80 has been reported for the time trade-off technique (Churchill *et al.* 1984b). The upper end of this range is quite acceptable. On the other hand, one-year test–retest r's have been poor for all techniques, ranging from 0.49 to 0.62 (Torrance 1976). However this may simply indicate, at least in part, that people's preferences shift over time. On balance then, the work to date suggests that over short intervals of time, the reliability of the measurement instruments is satisfactory.

Reliability can also be expressed in terms of the precision of an individual measurement, where σ_e is the standard deviation of the measurement error. For the rating scale σ_e values have been reported from 0.09 to 0.15 (Torrance *et al.* 1982), while values for the time trade-off and standard gamble are 0.13 (Torrance 1976). This means that if an individual responded to a standard gamble or a time trade-off question with a utility of 0.60 for a particular health state, the 95 per cent confidence interval would be 0.34 to 0.86, a rather wide range. Thus, single individual measurements are not particularly precise.

Moreover, individuals differed greatly in their health state preferences and the differences cannot be explained by the usual demographic characteristics such as age, sex, socioeconomic status, religion, illness, occupation, etc. For example, individual differences for the same health state on the 0–1 utility scale result in a standard deviation of scores of approximately 0.30 (Sackett and Torrance 1978).

Fortunately, the imprecision of individual measurements and the considerable differences among individuals can be ameliorated by taking the mean of a large group of subjects. Since the standard error of the mean is $0.30/\sqrt{N}$, the mean utility values for a health state can be made as precise as desired by increasing the group size N. Moreover, group mean values have been found to be remarkably stable regardless of the make-up of the group (Sackett and Torrance 1978; Boyd *et al.* 1982; Wolfson *et al.*, 1982). These are fortunate findings since group mean utilities are the values normally required in programme evaluation.

A measure is valid if it accurately reflects the concept or phenomenon it claims to measure. Two approaches have been taken here. In one approach, health state utilities are claimed to be utilities obeying the axioms of von Neumann–Morgenstern utility theory for decisions under uncertainty (von Neumann and Morgenstern 1953). In this case, the standard gamble measurement technique is valid by definition and the validity of the other techniques can be determined by comparison. In one study using this approach the time trade-off technique was found to be relatively valid while the rating scale method was not (Torrance 1976). In the other approach to validity, health state utilities are claimed to measure the overall quality of life associated with the health state and thus should be strongly correlated with other trusted measures of health-related quality of life. Using this approach, Churchill *et al.* (1984b) determined that patients' measured utilities correlate significantly with nephrologists' ratings of the patients' quality of life. Moreover, groups of patients with different clinical status differed in their measured health state utilities in predictable ways.

Given the uncertainties still surrounding the reliability and validity of utility values, it is important to perform sensitivity analyses on them. If wide changes in the utility values have no impact on the study's conclusion, as was the case in one recent study (Boyle, Torrance, Sinclair, and Horwood 1983), nothing further need be done. If, on the other hand, the conclusions of the study are found to be sensitive to the utility values for specific health states, these values should be remeasured on additional subjects, perhaps with other techniques, to improve the confidence in the overall result.

6.1.7. Is it all worthwhile?

In some of the foregoing material we have dealt rather extensively with potential measurement problems and biases, and the lack of definitive advice on how best to measure utilities. We believe that it is important for the reader to be aware of the potential hazards and pitfalls, in order to do good work. But please do not be discouraged. Our experience is that in practice these measurements are not as onerous as they may at first appear. And our conviction is that for quality economic appraisals these measurements are often essential—for it is far better to have an approximate measure of the right factors than a precise measure of the wrong ones.

6.2. UTILITY VALUES AVAILABLE FROM THE LITERATURE

There is a growing body of studies in which utilities have been measured for a few specific health states. For example, Sackett and Torrance (1978) presented utilities for depression, home confinement for tuberculosis, home confinement for an unnamed contagious disease, hospital confinement for tuberculosis, hospital confinement for an unnamed contagious disease, hospital dialysis, home dialysis, kidney transplant, mastectomy for breast cancer, and mastectomy for injury. Churchill *et al*. (1984b) reported utilities for hospital dialysis and continuous ambulatory peritoneal dialysis. Utilities for loss of speech due to laryngectomy have been reported by McNeil *et al*. (1981). Utilities for cancer-related states have been studied and reported by the group at Princess Margaret Hospital, Toronto (Llewellyn-Thomas, Sutherland, Tibshirani, Ciampi,

Cost-utility analysis

Till, and Boyd 1982; Sutherland, Dunn, and Boyd 1983). Pliskin *et al.* (1980) reported utilities for two levels of angina pain—mild and severe.

There are two studies in which a classification scheme covering a fairly wide range of health states has been established, and utility values have been measured for the states. One is the study by Bush and his colleagues reported in Kaplan *et al.* (1976), in which the measurements were performed using a rating scale technique on the general public in San Diego. The other is a McMaster study reported in Torrance *et al.* (1982) and summarized below, in which the measurements were performed using both rating scale and time trade-off measures within a multi-attribute utility theory framework.

In the study by Torrance *et al.* (1982), the four-attribute health state classification system shown in Table 6.1 was designed to uniquely categorize the status of all individuals two years of age and over. It was successfully applied to classify the outcomes of neonatal intensive care and to follow randomly selected children over time (Boyle *et al.* 1983). Not only were the children followed to their current age (≤15 years), but their future health pattern was forecast by two developmental paediatricians using the health state classification system. The system appears to be useful to classify the health states of children and adults for a wide variety of conditions.

Each of the four attributes in the health state classification system is subdivided into a number of levels. To classify the health state of an individual at a point in time, a level is selected on each attribute that best represents the level of functioning of the individual on that attribute. Thus the health state of the individual is specified by indicating four levels. For example, a perfectly healthy individual would be (P1,R1, S1, H1), a physically handicapped individual might be (P3, R2, S1,H1)—note that H includes only additional health problems not included in the other three attributes—and an emotionally disturbed individual might be (P1, R2, S4, H1).

There are 960 possible health states (all combinations of all levels). A utility value can be determined for each of these 960 health states by using the formula given below. The data for the formula come from utility measurements taken on a sample of healthy adults in the community. The model for the formula comes from curve A in Fig. 1 of the original work (Torrance *et al.* 1982). This is one of three models (curves A, B, and C) originally described, based on three different methodological assumptions.

Model A was selected for the formula because it tentatively appears, in application, to be the most valid. The formula itself is a slightly simplified version of model A, to facilitate calculations. The simplification provides

120

a very good fit over most of the range, and a slight positive bias for the very worst health states, those considered worse than death. However, since not much is known about the appropriate measurement of utility values of health states worse than death (Torrance 1984), this simplification is not likely to be much of a problem in practice.

In summary, the formula given here is based on our best data and knowledge to date, but undoubtedly it is not the last word. It can be very helpful as a simple and quick first approximation, particularly when coupled with sensitivity analysis.

1. Formula for utility values

The formula gives utility values on the standard scale where healthy is 1.00 and dead is 0.00. However, since some of the health states were judged to be worse than death, some of the utility values are less than zero. The least utility value, for health state (P6, R5, S4, H8), is -0.21.

$$U = 1.42 \, (m_1 m_2 m_3 m_4) - 0.42$$

where U = utility of health state; m_1 = multiplicative utility factor for the level on attribute 1 from Table 6.2; m_2 = multiplicative utility factor for the level on attribute 2 from Table 6.2, etc. Example calculations:

$$U(\text{P1, R1, S1, H1}) = 1.42(1.00 \times 1.00 \times 1.00 \times 1.00) - 0.42 = 1.00$$
$$U(\text{P1, R1, S1, H4}) = 1.42(1.00 \times 1.00 \times 1.00 \times 0.91) - 0.42 = 0.87$$
$$U(\text{P3, R2, S1, H1}) = 1.42(0.81 \times 0.94 \times 1.00 \times 1.00) - 0.42 = 0.66$$
$$U(\text{P1, R2, S4, H1}) = 1.42(1.00 \times 0.94 \times 0.77 \times 1.00) - 0.42 = 0.61$$
$$U(\text{P3, R2, S2, H5}) = 1.42(0.81 \times 0.94 \times 0.96 \times 0.86) - 0.42 = 0.47$$
$$U(\text{P5, R4, S3, H1}) = 1.42(0.61 \times 0.75 \times 0.86 \times 1.00) - 0.42 = 0.14$$
$$U(\text{P5, R5, S4, H7}) = 1.42(0.61 \times 0.50 \times 0.77 \times 0.83) - 0.42 = -0.14$$
$$U(\text{P6, R5, S4, H8}) = 1.42(0.52 \times 0.50 \times 0.77 \times 0.74) - 0.42 = -0.21$$

2. *Sensitivity analysis*

Although the formula produces a single utility value for each health state, the measurements on which the formula is based are not precise. Measurement uncertainty includes both sampling error and measurement imprecision. These are combined in the standard error $S_{\bar{x}} = 0.06$. A sensitivity analysis of $\pm 2 S_{\bar{x}}$ would give an upper bound utility value of $U + 0.12$, not to exceed 1.00, and a lower bound of $U - 0.12$.

3. *Conclusion*

The health state classification system and utility formula presented here can be used in many studies as, at least, an initial method of assigning utility values to health outcomes.

Table 6.1. Health state classification system (Age >2 years)

X₁ Physical function: mobility and physical activity[a]

Level x_1	Code	Description
1	P1	Being able to get around the house, yard, neighbourhood or community WITHOUT HELP from another person; AND having NO limitation in physical ability to lift, walk, run, jump or bend.
2	P2	Being able to get around the house, yard, neighbourhood or community WITHOUT HELP from another person; AND having SOME limitations in physical ability to lift, walk, run, jump or bend.
3	P3	Being able to get around the house, yard, neighbourhood or community WITHOUT HELP from another person; AND NEEDING mechanical aids to walk or get around.
4	P4	NEEDING HELP from another person in order to get around the house, yard, neighbourhood or community; AND having SOME limitations in physical ability to lift, walk, run, jump or bend.
5	P5	NEEDING HELP from another person in order to get around the house, yard, neighbourhood or community; AND NEEDING mechanical aids to walk or get around.
6	P6	NEEDING HELP from another person in order to get around the house, yard, neighbourhood or community; AND NOT being able to use or control the arms and legs.

X₂ Role function: self-care and role activity[a]

Level x_2	Code	Description
1	R1	Being able to eat, dress, bathe, and go to the toilet WITHOUT HELP; AND having NO limitations when playing, going to school, working or in other activities.
2	R2	Being able to eat, dress, bathe and go to the toilet WITHOUT HELP; AND having SOME limitations when playing, going to school, working or in other activities.
3	R3	Being able to eat, dress, bathe and go to the toilet WITHOUT HELP; AND NOT being able to play, go to school or work.

| 4 | R4 | NEEDING HELP to eat, dress, bathe or go to the toilet; AND having SOME limitations when playing, going to school, working or in other activities. |
| 5 | R5 | NEEDING HELP to eat, dress, bathe or go to the toilet; AND NOT being able to play, attend school or work. |

X_3 Social-emotional function: emotional wellbeing and social activity

Level x_3	Code	Description
1	S1	Being happy and relaxed most or all of the time, AND having an average number of friends and contacts with others.
2	S2	Being happy and relaxed most or all of the time, AND having very few friends and little contact with others.
3	S3	Being anxious or depressed some or a good bit of the time, AND having an average number of friends and contacts with others
4	S4	Being anxious or depressed some or a good bit of the time, AND having very few friends and little contact with others.

X_4 Health problem[b]

Level x_4	Code	Description
1	H1	Having no health problem.
2	H2	Having a minor physical deformity or disfigurement such as scars on the face.
3	H3	Needing a hearing aid.
4	H4	Having a medical problem which causes pain or discomfort for a few days in a row every two months.
5	H5	Needing to go to a special school because of trouble learning or remembering things.
6	H6	Having trouble seeing even when wearing glasses.
7	H7	Having trouble being understood by others.
8	H8	Being blind OR deaf OR not able to speak.

[a] Multiple choices within each description are applied to individuals as appropriate for their age. For example, a 3-year-old child is not expected to be able to get around the community without help from another person.

[b] Individuals with more than one health problem are classified according to the problem they consider the most serious.

Table 6.2. Multiplicative utility factors

Physical function		Role function		Social–emotional function		Health problem	
Level	Multiplicative Utility factor m_1	Level	Multiplicative Utility factor m_2	Level	Multiplicative Utility factor m_3	Level	Multiplicative Utility factor m_4
P1	1.00	R1	1.00	S1	1.00	H1	1.00
P2	0.91	R2	0.94	S2	0.96	H2	0.92
P3	0.81	R3	0.77	S3	0.86	H3	0.91
P4	0.80	R4	0.75	S4	0.77	H4	0.91
P5	0.61	R5	0.50			H5	0.86
P6	0.52					H6	0.84
						H7	0.83
						H8	0.74

6.3 MEASURING UTILITIES: METHODS AND AN EXERCISE BASED ON A SIMULATED INTERVIEW

A detailed description of the methods of measuring health state utilities is available in Torrance (1982). A brief summary of these methods is provided here.

Two of the first tasks are to identify and to describe the relevant health states for which utilities are required, and to determine the subjects on whom the utility measurements are to be made. These steps were discussed earlier in Section 6.1.5. There are three commonly used methods of measuring cardinal utilities of health states: rating scale, standard gamble and time trade-off.

6.3.1. Rating scale

A typical rating scale consists of a line on a page with clearly defined endpoints. The most preferred health state is placed at one end of the line and the least preferred at the other end. The remaining health states are placed on the line between these two, in order of their preference, and such that the intervals or spacing between the placements correspond to the differences in preference as perceived by the subject.

Preferences for chronic states can be measured on a rating scale. The chronic states are described to the subject as irreversible; that is, they are to be considered permanent from age of onset until death. The subject must be provided with the age of onset and the age of death, and these should be the same for all states that are measured together relative to each other in one batch. States with different ages of onset and/or ages of death can be handled by using multiple batches. Two additional chronic states are added to each batch as reference states for the scale—healthy (from age of onset to age of death) and death (at age of onset).

The subject is asked to select the best health state of the batch, which presumably would be 'normal healthy life' and the worst state, which may or may not be 'death at age of onset'. He is then asked to locate the other states on the rating scale relative to each other such that the distances between the locations are proportional to his preference differences. The rating scale is measured between 0 at one end and 1 at the other end. If death is judged to be the worst state and placed at the 0 end of the rating scale, the preference value for each of the other states is simply the scale value of its placement. If death is not judged to be the worst state but is placed at some intermediate point on the scale, say d, the preference values for the other states are given by the formula $(x - d)/(1 - d)$, where x is the scale placement of the health state.

Cost-utility analysis

Preferences for temporary health states can also be measured on a rating scale. Temporary states are described to the subject as lasting for a specified duration of time at the end of which the person returns to normal health. As with chronic states, temporary states of the same duration and same age of onset should be batched together for measurement. Each batch should have one additional state 'healthy' added to it. The subject is then asked to place the best state (healthy) at one end of the scale and the worst temporary state at the other end. The remaining temporary states are located on the scale such that the distances between the locations are proportional to his preference differences.

If the programmes being evaluated involve only morbidity and not mortality and if there is no need to compare the findings to programmes that do involve mortality, the procedure described above for temporary health states is sufficient. However, if this is not the case, the interval preference values for the temporary states must be transformed onto the standard 0–1 health preference scale. This can be done by redefining the worst temporary health state as a chronic state of the same duration, and measuring its preference value by the technique described for chronic states. The values for the other temporary health states can then be transformed onto the standard 0–1 dead–healthy scale by a positive linear transformation (just like converting °F to °C).

6.3.2. Standard gamble

The standard gamble is the classical method of measuring cardinal preferences. It is based directly on the fundamental axioms of utility theory, first presented by von Neumann and Morganstern (1953). The method has been used extensively in the field of decision analysis, and good descriptions of the methods are available in books in this field (for example see Holloway 1979).

The method can be used to measure preferences for chronic states but the method varies somewhat depending upon whether or not the chronic state is preferred to death. For chronic states preferred to death the method is displayed in Fig. 6.1. The subject is offered two alternatives. Alternative 1 is a treatment with two possible outcomes: either the patient is returned to normal health and lives for an additional t years (probability p), or the patient dies immediately (probability $1 - p$). Alternative 2 has the certain outcome of chronic state i for life (t years). Probability p is varied until the respondent is indifferent between the two alternatives, at which point the required preference value for state i is simply p; that is, $h_i = p$.

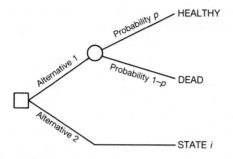

Fig. 6.1. Standard gamble for a chronic health state preferred to death.

Since most subjects cannot readily relate to probabilities, the standard gamble is often supplemented with the use of visual aids, particularly a probability wheel (Torrance 1976). This is an adjustable disk with two sectors, each of different colour, and constructed so that the relative size of the two sectors can be readily changed. The alternatives are displayed to the subject on cards, and the two outcomes of the gamble alternative are colour-keyed to the two sectors of the probability wheel. The subject is told that the chance of each outcome is proportional to the similarly coloured area of the disk.

Preferences for temporary health states can be measured relative to each other using the standard gamble method as shown in Fig. 6.2. Here intermediate states i are measured relative to the best state (healthy) and the worst state (temporary state j). In this format the formula is $h_i = p + (1 - p)h_j$ where i is the state being measured and j is the worst

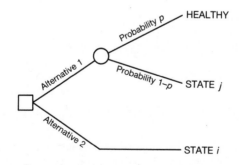

Fig. 6.2. Standard gamble for a temporary health state.

state. If death is not a consideration in the use of the utilities, h_j can be set equal to zero and the h_i values determined from the formula which then reduces to $h_i = p$. However, if it is desired to relate these values to the 0–1 dead–healthy scale, the worst of the temporary states (state j) must be redefined as a short duration chronic state (followed by death) and measured on the 0–1 scale by the technique described above for chronic states. This gives the value for h_j which can then, in turn, be used in the above formula to find the value for the h_i.

Variations on this method are also possible. For example, in Fig. 6.2 state j can be the state immediately dispreferred to state i, rather than the worst state. This does not change the formula $h_i = p + (1 - p)h_j$ but it does mean that the h values for the states have to be solved in sequence, starting with the worst state. This variation is used in the simulated interview in this section.

6.3.3. Time trade-off

The time trade-off method was developed specifically for use in health care by Torrance, Thomas, and Sackett (1972). The application of the time trade-off technique to a chronic state considered better than death is shown in Fig. 6.3. The subject is offered two alternatives:

1. state i for time t (life expectancy of an individual with the chronic condition) followed by death;
2. healthy for time $x < t$ followed by death.

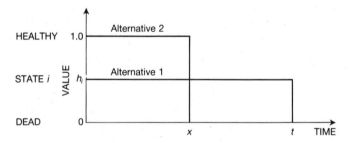

Fig. 6.3. Time trade-off for a chronic health state preferred to death.

Time x is varied until the respondent is indifferent between the two alternatives, at which point the required preference value for state i is given by $h_i = x/t$.

Preferences for temporary health states can be measured relative to each other using the time trade-off method as shown in Fig. 6.4. As with the rating scale and the standard gamble, intermediate states i are measured relative to the best state (healthy) and the worst state (temporary state j). The subject is offered two alternatives:

1. temporary state i for time t (the time duration specified for the temporary states), followed by healthy;
2. temporary state j for time $x < t$, followed by healthy.

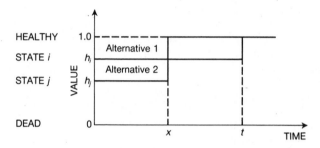

Fig. 6.4. Time trade-off for a temporary health state.

Time x is varied until the respondent is indifferent between the two alternatives, at which point the required preference value for state i is $h_i = 1 - (1 - h_j)x/t$. If we set $h_j = 0$, this reduces to $h_i = 1 - x/t$. Figure 6.4 shows the basic format, but other variations are possible. State j need not be the worst state as long as it is any state worse than i. In using variations, however, care must be taken to ensure that all preference values can be calculated. In one systematic variation which has been used (Torrance *et al.* 1972; Torrance 1976; Sackett and Torrance 1978), state j is always the next worse state to state i. This variation is used in the simulated interview in this section. Although the formula is still the same, $h_i = 1 - (1 - h_j)x/t$, the states must now be solved in sequence from worst to best.

Finally, as with the rating scale and the standard gamble, if the preference values for the temporary states are to be transformed to the 0–1 dead–healthy scale, the worst of the temporary states must be redefined as a short duration chronic state and measured by the method for chronic states described above.

6.3.4. Simulated interview

The following example and exercise is provided to give some feeling for how these instruments are actually used in practice.

Consider the following simulated interview between an interviewer (I) and a subject (S) to measure the utility of three chronic health states (A, B, C) and three temporary states (D, E, F). The utility scale is the conventional one anchored by dead = 0 and healthy = 1. Each state is measured three times, by three different techniques: standard gamble, time trade-off and category scaling. The task is to determine eighteen utility values (six states, three techniques). (Answers in Table 6.3.)

Hint: In the standard gamble and the time trade-off methods you must determine the subject's 'indifference point'. This is the point at which the subject is indifferent between (cannot decide between) choice 1 and choice 2. If the subject switches her choice on two adjacent questions, the indifference point is taken to be halfway between. If the subject expresses indifference on a particular question, that is taken as the indifference point.

Notation: I = Interviewer
S = Subject
A > B means A *is preferred to* B

1. *Chronic health states*

I: Thank you for agreeing to participate.
On each of these three sheets is a description of a chronic condition. Each sheet is labelled A, B, or C. [*Note to reader*: The conditions can be quite disparate, like kidney dialysis, blind, mental retardation.] Plese imagine that you will have to spend the rest of your life in one of these conditions. Rank the conditions in order of preference, and relative to 'healthy' and 'dead'. That is, if any of the conditions are better than being healthy or worse than being dead, please indicate.
S: Alright, healthy > A > B > C > dead.
I: Good. Now I have a device here called a probability wheel.
[*References*: Torrance 1976, p. 131; Torrance 1982, p. 41.] The wheel is divided into two sectors; a blue sector and a yellow sector. First let us adjust the wheel so it is half blue and half yellow. Now consider the following choices; which do you prefer, choice 1 or choice 2? In Choice 1 you get chronic condition C for the rest of your life. In Choice 2 you get either 2a or 2b described below depending on the outcome of the spin of this pointer on the wheel. That is, if you choose 2, I am going to spin this pointer on the wheel and if it stops on

blue you get 2a but if it stops on yellow you get 2b. 2a is healthy for the rest of your life. 2b is immediate death. Now which do you choose, 1 or 2?

S: Well C is pretty bad, I think I'd take my chances and select Choice 2.

I: Alright. Now I am going to adjust the wheel so it is only 40 per cent blue and 60 per cent yellow. Now the pointer has a greater chance of stopping on yellow in which case you get 2b, immediate death. Would you still select Choice 2.

S: Yes.

I: O.K., now I'll adjust the wheel again so it is 30 per cent blue, 70 per cent yellow. Would you still select Choice 2?

S: No, I don't think so. That's too risky now, I'd take Choice 1.

I: O.K., that's the end of that one, thank you. Now, imagine a new situation. Here you again have two choices, and I'll want to know which you prefer. Choice 1 is condition B for the rest of your life. Choice 2 is the same as before. Let's reset the wheel to 50 per cent blue and 50 per cent yellow. Now, which choice do you prefer?

S: I think I'd take Choice 1.

I: O.K., now I'll adjust the wheel to 60 per cent blue, 40 per cent yellow. Now, how do you feel about the choices?

S: Oh, that makes Choice 2 more attractive. I think I'll switch to it.

I: Fine, thank you. Now we have another new situation. This time Choice 1 is condition A for the rest of your life. Choice 2 is the same as before. Let's reset the wheel to 50/50, now which do you prefer?

S: Choice 1 definitely.

I: O.K., now I'll adjust the wheel to 60 per cent blue, 40 per cent yellow. Now what do you think?

S: Still Choice 1.

I: O.K., let's move the wheel to 70 per cent blue, and 30 per cent yellow.

S: Still Choice 1.

I: O.K., let's move the wheel to 80 per cent blue, 20 per cent yellow.

S: Now, that makes it difficult to decide. I think Choice 1 and Choice 2 now seem about the same to me. I really don't much care which I get.

I: O.K., fine thank you. Now I have a new type of question for you. Given your age, your remaining life expectancy is another 40 years. I am going to give you some choices and ask you which you prefer. Choice 1 is to live your remaining 40 years in condition C. Choice 2 is to live a shorter time, but healthy. For example, let's say in Choice 2 you get to live only 20 more years, but healthy. Which would you take, Choice 1 or Choice 2?

S: I'd say Choice 2, because C is really pretty bad.

I: O.K., what if Choice 2 was only 15 years, but still healthy.

S: Oh, that's getting pretty tough, but I think I'd still take Choice 2.

I: O.K., what if choice 2 is only 14 years?

S: That's about my limit. Any less than 14 years I'll switch to Choice 1. Fourteen years is right on the knife edge, I could go either way.

I: O.K., fine thank you. Now a new situation. Choice 1 is condition B for your remaining 40 years. In Choice 2 you will have only 20 years to live, but they will be healthy years.

S: Well that's pretty hard to choose. I think those two choices are about the same.

I: Fine. Now for the last situation concerning these chronic states. Choice 1 is condition A for your remaining 40 years. In Choice 2 you will have only 20 years to live but they will be healthy years.

S: Well, I think this time I would take Choice 1.

I: O.K., let's make Choice 2, 25 years, all healthy.

S: I'd still take Choice 1.

I: O.K., let's make Choice 2, 30 years, all healthy.

S: I'd still take Choice 1.

I: O.K., let's make Choice 2, 35 years, all healthy.

S: O.K., this time I would take Choice 2.

I: Fine, thank you. Now for the last type of question relating to these health states, I have here a device we call a 'feeling thermometer' [*see* Torrance 1982; p. 40]. As you can see it is a thermometer-shaped 0–100 scale on a felt board, with 0 labelled 'least desirable' and 100 labelled 'most desirable'. I also have these five narrow foam sticks pointed at each end labelled respectively healthy, dead, A, B, and C. I want you to select the foam stick that represents the most desirable way to spend the rest of your life and place it on the felt board beside the thermometer at 100. Similarly, please select the least desirable and place it at 0. Now place the other sticks somewhere between the top and the bottom of the thermometer depending on how you feel about spending the rest of your life in each condition. If you feel the same about two sticks just put them on top of each other. The distance between the sticks should all be relative to each other. For example, if you feel that the difference in desirability between healthy and A is twice as great as the difference between A and B, the spacing between the sticks healthy and A should be adjusted to be twice that between A and B.

S: O.K., I think that's it.

I: Thinking it over, are there any changes you would like to make?

S: No, I think not.

I: Fine, now please read out the sticks and the thermometer values they are beside.

S: O.K., Healthy 100, A 80, B 52, C 35, Dead 0.

I: Fine, thank you very much, that completes this phase of the interview.

2. *Temporary health states:*

I: Here are three different sheets. Each sheet describes a temporary dysfunctional health condition which you would have for three months. At the end of the three months you will be completely recovered. Each sheet is labelled D, E or F [*Note to reader*: The conditions can be quite disparate like mononucleosis for 3 months, hospital confinement with kidney dialysis for 3 months for the treatment of temporary kidney failure, clinical depression for 3 months.] Imagine that you will have to spend the next three months in one of these conditions. Please rank them in order of preference.

S: O.K., I'd say D > E > F.

I: Good, now I'm going to get out the probability wheel and give you some choices again. For this first situation imagine that, unfortunately, you have only 3 months to live. Choice 1 is to spend these 3 months in condition F. In Choice 2 you get either 2a or 2b depending on the spin of the pointer on the wheel; blue gives you 2a, yellow 2b: 2a is healthy for the 3 months, 2b is immediate death. With the wheel at 50 per cent blue, 50 per cent yellow which would you choose?

S: F is pretty bad, I'll take Choice 2.

I: O.K., 40 per cent blue, 60 per cent yellow.

S: That's about my indifference point. Any less blue and I'll switch for sure.

I: Fine, now here's a completely new situation, that fortunately does not involve dying. This time you will always completely recover at the end of the 3 months. Choice 1 is condition E for the 3 months. Choice 2 is either 2a (if blue) or 2b (if yellow) depending on the outcome of the spin. 2a is immediate cure; 2b is condition F for 3 months. With the wheel at 50 per cent blue, 50 per cent yellow which would you choose?

S: Choice 1.

I: O.K., 60 per cent blue, 40 per cent yellow.

S: Choice 2.

I: Good. Thanks. Now for a similar question.

Choice 1 is condition D for 3 months followed by cure. Choice 2, is a spin leading to: 2a (blue), immediate cure; 2b (yellow), condition E for 3 months followed by cure.

With the wheel at 50/50 which would you choose?

S: Choice 2.

I: O.K., 40 per cent blue, 60 per cent yellow?

S: Still Choice 2.

I: O.K., 30 per cent blue, 70 per cent yellow?

S: O.K., now I'll switch to Choice 1.

I: Good, thanks. That completes the probability wheel questions. Now, here's a new question. Unfortunately, in this question we must again assume you have only 3 months to live. Choice 1 is to spend this last 12 weeks in condition F. In Choice 2 you live for a shorter period of time, but healthy. Let's say 6 weeks.

S: I'd take Choice 2.

I: O.K., what if it was only 5 weeks?

S: That's about my indifference point.

I: Good, now we'll change the situation so you no longer die. Now, you will be completely cured at the end of the temporary condition. Which would you prefer: Choice 1, 12 weeks of E; Choice 2, 6 weeks of F?

S: Choice 1.

I: O.K., what if Choice 2 is only 5 weeks?

S: O.K., now I'd take Choice 2.

I: Good, now for the last situation. Again the choices are followed by complete cure. Choice 1, 12 weeks of D; Choice 2, 6 weeks of E.

S: Choice 2.

I: O.K., what if Choice 2 is 7 weeks?

S: I'd still take it.

I: What if it is 8 weeks?

S: Now I can't tell. That's my indifference point. Any longer than that and I'd switch to Choice 1.

I: Good, that completes these questions. Now we just have the feeling thermometer to do again and then we're finished. This time the five sticks are labelled healthy, dead, D, E, and F. Furthermore, this time we must assume that you have only three months to live. Please place the five sticks on the feeling thermometer following the same instructions as the last time. When you're ready please read them out to me.

S: O.K., here they come. Healthy 100, D 81, E 75, F 41, Dead 0.

I: O.K., terrific. That completes the interview. Thank you very much. You have been very helpful.

6.4. CALCULATIONS OF HEALTH IMPROVEMENT USING UTILITIES

The following examples show how to calculate quality-adjusted life-years (QALY) gained with and without discounting, given the health state utilities and the programme effectiveness data. The utilities are taken from Sackett and Torrance (1978) Table 2, reproduced here as Table 6.4. The discounting calculations use the tabled factors consistent with the discounting formulae given in Chapter 4, p. 48.

Table 6.3. Calculations of health state utilities for the simulated interview

	Standard gamble	Time trade-off	Rating scale (thermometer)
Chronic states			
lifetime			
Healthy	1.00	1.00	1.00
A	0.80	0.81	0.80
B	0.55	0.50	0.52
C	0.35	0.35	0.35
Dead	0.00	0.00	0.00
Temporary states			
3 months			
Healthy	1.00	1.00	1.00
D	0.82	0.82	0.81
E	0.73	0.73	0.75
F	0.40	0.42	0.41
Dead	0.00	0.00	0.00

Calculations

$C = 14/40 = 0.35$
$B = 20/40 = 0.50$
$A = 32.5/40 = 0.81$

$F = 0.40$
$E = 0.55 \times 1.00 + 0.45 \times 0.40 = 0.73$
$D = 0.35 \times 1.00 + 0.65 \times 0.73 = 0.82$

$F = 5/12 = 0.42$
$E = 1 - (1-0.42)(5.5/12) = 0.73$
$D = 1 - (1-0.73)(8/12) = 0.82$

Table 6.4. Mean daily health state utilities in the general population sample

Duration	Health state	Observations		Mean daily health state utility	Standard error
		Total	Useable		
	Reference state: Perfect health			1.00	
3 months	Home confinement for tuberculosis	246	239	0.68	0.020
3 months	Home confinement for an unnamed contagious disease	246	240	0.65	0.022
3 months	Hospital dialysis	246	243	0.62	0.023
3 months	Hospital confinement for tuberculosis	246	241	0.60	0.022
3 months	Hospital confinement for an unnamed contagious disease	246	242	0.56	0.023
3 months	Depression	246	243	0.44	0.024

8 years	Home dialysis	246	240	0.65	0.018
8 years	Mastectomy for injury	60[a]	56	0.63	0.038
8 years	Kidney transplant	246	242	0.58	0.021
8 years	Hospital dialysis	246	240	0.56	0.019
8 years	Mastectomy for breast cancer	60[a]	58	0.48	0.044
8 years	Hospital confinement for an unnamed contagious disease	246	241	0.33	0.022
Life	Home dialysis	197[b]	187	0.40	0.031
Life	Hospital dialysis	197[b]	189	0.32	0.028
Life	Hospital confinement for an unnamed contagious disease	197[b]	192	0.16	0.020
	Reference state: dead			0.00	
	Total	*3171*	*3093*		

[a] This scenario was presented to 41–79 year old females only.
[b] This scenario was presented to 18–65 year old subjects only.

Cost-utility analysis

6.4.1. Examples

1. How many *Q*uality-*A*djusted-*L*ife-*Y*ears (QALYs) are gained if a person achieves an 8-year life extension on home dialysis,

   (a) assuming no discounting
   (b) assuming discounting at a rate of 10 per cent per annum?*

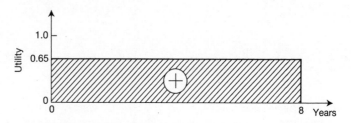

   (a) $0.65 \times 8 = 5.2$ QALYs
   (b) $0.65 \times (4.8684 + 1.0000) = 3.8$ QALYs

2. How many QALYs are gained if a person achieves a 3-month life extension on hospital dialysis,

   (a) assuming no discounting?
   (b) assuming discounting at 10 per cent per annum?*

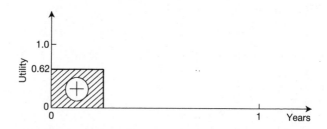

   (a) $0.62 \times 0.25 = 0.16$ QALY
   (b) $0.62 \times 0.25 = 0.16$ QALY

3. How many QALYs are gained by preventing a case of tuberculosis which would have been treated at home for three months,

   (a) assuming no discounting?
   (b) assuming discounting at 10 per cent?

* For discounting purposes assume that all health gains or losses that occur throughout a year take place at the beginning of the year.

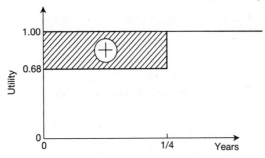

Cost-utility analysis

(a) $(1.00 - 0.68) \times 1/4 = 0.32 \times 1/4 = 0.08$ QALY
(b) $(1.00 - 0.68) \times 1/4 = 0.32 \times 1/4 = 0.08$ QALY

4. Assume a breast cancer patient will become symptomatic, have a mastectomy and live an additional six years. By screening, you can detect the breast cancer one year earlier, perform the mastectomy one year earlier and add two years to the patient's life (that is, she now lives nine years from the mastectomy instead of six). How many QALYs are gained by screening,

    (a) assuming no discounting?
    (b) assuming discounting at 10 per cent?

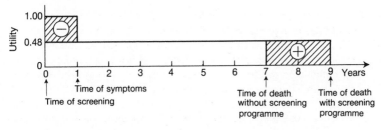

(a) $0.48 \times 2 - (1 - 0.48) \times 1 = 0.96 - 0.52 = 0.44$ QALY
(b) $0.48 \times 0.5132 + 0.48 \times 0.4665 - 0.52 \times 1.000 = 0.47 - 0.52$
    $= -0.05$ QALY

6.5. CRITICAL APPRAISAL OF A PUBLISHED ARTICLE

Reference: Boyle, M. H., Torrance, G. W., Sinclair, J. C. and Horwood, S. P. (1983). Economic evaluation of neonatal intensive care of very-low-birth-weight infants. *N. Engl. J. Med.* **308**, 1330–7.

Cost-utility analysis

This paper is assessed below using the 10 questions set out in Annex 3.1. It is suggested that you locate the article and attempt the exercise before reading the assessment.

1. Was a well-defined question posed in answerable form?

__x__ YES _____ NO _____ CAN'T TELL

The study examines both the costs and health outcomes of management of very-low-birth-weight babies either by a regional programme with a specialized neonatal intensive care unit or by previously existing facilities within the region.

The viewpoint for the analysis is clearly that of society since, as the authors explain, '... we measured all the costs and benefits of providing neonatal intensive care, *regardless of who pays or who benefits*' (p. 1330 of the paper).

2. Was a comprehensive description of the competing alternatives given?

__x__ YES _____ NO _____ CAN'T TELL

The intervention under study is the neonatal intensive care component of the McMaster regional perinatal programme. This intervention, therefore, consists of both specific clinical services (the neonatal intensive care unit itself and the specialized procedures performed therein) and the programmatic aspects of the management of very-low-birth-weight babies within the region (consultation, referral, transport, etc.). A general description of the setting, population served, levels of care provided, frequency of neonatal intensive care, and hospitals involved is provided on pp. 1330–1. For more detailed description, readers are provided with references to recommendations for the regional development of perinatal health services (p. 1331, col. 1) to which this programme conformed, and with a reference to an earlier article by the same authors [Evaluation of neonatal intensive care programs. *N. Engl. J. Med.* (1981) **305**, 489–94] which identified several specific clinical manœuvres typically included in a neonatal intensive care programme (Table 1, p. 490). The neonatal intensive care 'programme' is compared to the 'no programme' situation in which care was previously delivered in individual hospitals within the region in the absence of a specialized neonatal intensive care unit at McMaster and an organized referral network.

140

3. Was there evidence that the programmes' effectiveness had been established?

___x___ YES _____ NO _____ CAN'T TELL

This assessment might be questioned; however, in their previous article, cited above, the authors reviewed the evidence on efficacy and effectiveness of several practices involving the respiratory, nutritional and environmental management of sickness in infants. They found experimental evidence supporting the efficacy of many of the specific manœuvres typically included in a neonatal intensive care programme. Yet, they also emphasized that methodological difficulties clouded conclusions on whether regional neonatal intensive care programmes were effective (p. 490).

In their economic evaluation of the neonatal intensive care (NIC) component of the McMaster regional perinatal programme, the effectiveness of the intervention is evaluated simultaneously with its efficiency through a 'before–after' research design. The health outcomes of all live-born infants weighing 500–1499 g, born to residents of Hamilton–Wentworth County, were compared for two groups: infants born from 1964–1969 (before the NIC programme) and infants born from 1973–1977 (after the introduction of NIC). Rate of survival to hospital discharge, life years per live birth and quality-adjusted life years per live birth all increased with NIC, thereby indicating programme effectiveness.

The before–after design is, of course, a less powerful research design than the randomized controlled trial. There was no concurrent control group and a major concern is whether the comparability of the two groups on factors other than NIC may have influenced infant outcomes. For example, there may have been differences between the groups in neonatal mortality risk. The authors were clearly aware of the limitations of the research design that they were forced to use. They performed such analyses as were possible in order to minimize potential bias, and identified potential confounding variables. For example, as noted on p. 1334, col. 2, they carried out birth-weight specific comparisons to minimize confounding from factors such as maternal age, parity and socioeconomic status known to affect the birth-weight distribution of live births. They also ascertained that other factors, capable of influencing birth-weight specific mortality risk at the time of birth (e.g., mode of delivery), did not represent an important difference between the groups.

4. Were all the important and relevant costs and consequences for each alternative identified?

___x___ YES _____ NO _____ CAN'T TELL

With respect to the costs of organizing and operating NIC, the authors identified hospital costs (including both capital and labour costs generated in the NIC unit and various wards, support services such as housekeeping and overhead services such as administration), physician services, convalescent care in community hospitals outside Hamilton, and ambulance transport (p. 1331, col. 2). They also identified as 'follow-up' costs the costs of caring for infants postdischarge, including both items such as hospital and physician services/dental services/drugs and items such as special institutional care applied and the extra cost of special education (p. 1331, col. 2).

(Note that although the authors label these follow-up items as 'costs' of the programmes, the items could have been treated as 'consequences' of the programmes. As consequences, they would have counted as 'changes in future resource use' and any savings here would have appeared as a direct benefit in later calculations. Although some intermediate calculations would have shown different absolute figures, any calculation of net economic benefit would not have changed and the conclusions would, of course, be unaffected by this alternative formulation.)

The authors also identified important intangible and emotional costs and consequences of the programme (p. 1335, col. 1) although these items were not measurable. While they did not include direct or indirect costs incurred by families up to the point of discharge, they did identify and impute a value to the provision of care to handicapped children by parents in their homes.

The consequences of the alternatives included both their health outcomes and the ensuing quality of life for survivors. In addition, the impact on the ability of the survivors to eventually perform productive economic roles was included. Thus a comprehensive range of costs and consequences was identified, consistent with the adoption of a societal viewpoint.

5. Were costs and consequences measured accurately in appropriate physical units?

___x___ YES _____ NO _____ CAN'T TELL

In order to measure hospital services consumed, the authors first identified, in specific measurement units, the different types of service which might be utilized during an episode of care (e.g., patient days for nursing and 'hotel' services, work units for radiology, operations for surgical services, etc.) and then recorded through chart review the actual utilization of services by infants in the study. Actual utilization of physician services, convalescent care and ambulance services was also obtained from appropriate records (p. 1331, col. 2). Types and quantities of health care and other services utilized by surviving children were obtained in a home interview with a random sample of families (p. 1331, col. 2). As the authors noted on p. 1334, col. 2, the direction of bias resulting from parental recall is uncertain.

Programme effects were measured initially by lives saved, obtained from hospital discharge data. However, effectiveness was also measured by life-years gained, both to age 15 and to death. In addition, what the authors term the 'social value', also called the 'utility' of the programme consequences, was measured by quality-adjusted life-years gained, again both to age 15 and to death.

An important element in the physical measurement of both costs and effects was the estimation of both lifetime health outcomes and utilization of services for surviving children. The authors employed their own data (based on a sophisticated, multi-attribute, classification system for health states employed in the home interview) to make the estimates up to age 15 where necessary. They employed forecasts made independently by two developmental paediatricians using the available health history for each child to make the projections past age 15 (p. 1131). In order to reflect uncertainty about the future, probability distributions of outcomes rather than point estimates were made by the paediatricians, and the authors later conducted sensitivity analyses on life expectancies.

6. Were costs and consequences valued credibly?

___x___ YES _____ NO _____ CAN'T TELL

Costs and benefits were valued in 1978 Canadian dollars. State-of-the-art methods were employed in two instances—the valuation of hospital neonatal care services and the establishment of utility values for the complex array of possible health states for surviving children.

In order to cost NIC, the authors established a 'fully allocated unit price' (cost) for each of the services previously identified as potentially contributing to an episode of hospital care. This involved the use of a mathematical model which employed hospital budget data and cost allocation criteria to apportion an appropriate share of the expense of all relevant hospital departments to services used in NIC. Reference is provided to a separate publication containing the exact data and detailed methods employed (p. 1331). (A key feature of this methodology is that it not only provides a method for sorting out situations of joint use of resources by more than one ward or unit,but it also sorts out situations in which two hospital departments simultaneously provide services to each other.) The cost of an NIC episode was then calculated by summing the quantities of each service used, multiplied by its unit price.

The costs of other services used were established from actual records of charges or operating expenses, the sources of which are noted by the authors. In the case of costs for the future care of handicapped children at home or in institutions, estimates were based on previous studies or obtained from the responsible paying agencies (p. 1332, col. 1). Future health care cost estimates were based on age and sex-specific normal utilization patterns adjusted for children in this study by health states. The value of their economic productivity was based on official census data on normal age and sex-specific earnings patterns, adjusted by forecasts of their productive ability as a percentage of normal ability (pp. 1331–1332).

The valuation of life years gained by surviving children in terms of their quality was accomplished by multiplying time spent or forecasted to be spent in any specific health state by the utility value of that health state. The utility values were based upon the preferences of a local random sample of parents with school children (p. 1331). Note that these utility values were based on actual measurement of preferences rather than an arbitrary assignment of values, and included parents with and without normal children (which is consistent with a societal viewpoint). The authors refer readers interested in more details of the specific multi-attribute utility theory approach (employed to deal with the complex range of health outcomes) to a separate publication (p. 1331, col. 2, note 12).

As a result of these identification, measurement and valuation procedures, the authors were in a position to conduct any one or all of the cost-effectiveness, cost–benefit or cost–utility analyses.

7. Were costs and consequences adjusted for differential timing?

__x__ YES _____ NO _____ CAN'T TELL

A discount rate of 5 per cent per annum was applied to costs, earnings and effects occurring in the future in order to convert the future values to their equivalent present value. Although no rationale was provided for the choice of a 5 per cent discount rate,the authors considered values from 0 to 10 per cent in the sensitivity analysis—a range which is likely to encompass the rates typically employed in economic evaluations (and recommended by most governments).

It is interesting to note the impact that discounting has on the results of this study. Tables 3 and 4 differ only in that, in the latter, all future costs and effects have been discounted to their present value. The first line of both tables is identical since all costs and effects measured to the point of hospital discharge occur within year one. However, as the analysis is extended to age 15 of the survivors and to death, the effects of discounting become apparent. As the authors explain, 'Since neonatal intensive care has high initial costs in order to achieve later gains in numbers of life-years, numbers of quality-adjusted life-years, and productivity, discounting affects the later gains more than the initial costs and adversely affects the measures of economic evaluation' (p. 1333).

8. Was an incremental analysis of costs and consequences of alternatives performed?

__x__ YES _____ NO _____ CAN'T TELL

Although the costs and outcomes of the two alternative programmes are calculated separately, the results of the study are presented in incremental form, that is, as the *additional* cost of the NIC programme per *additional* effect (e.g., life saved, life-year gained, QALY gained) it achieves. Details of how the incremental data were calculated are provided in the appendix. As well, the appendix demonstrates that all of the data used to derive the efficiency estimates were introduced into the calculations in their incremental form.

9. Was a sensitivity analysis performed?

__x__ YES _____ NO _____ CAN'T TELL

Four factors were chosen for testing in the sensitivity analysis: the discount rate, life expectancy, condition of those lost to follow-up and utility values. Of these four, varying the discount rate appears to have the greatest impact on the results (see Fig. 2, p. 1325). In fact, the sensitivity analysis demonstrates that the choice of discount rate can be pivotal for at least one of the indicators of efficiency in one birth-weight group. The net economic benefit per live birth for the 1000–1499 g birth-weight group was positive at discount rates lower than approximately 3.5 per cent but negative at higher discount rates.

This study demonstrates that sensitivity analyses can also be used to increase confidence in the original results. For example, the authors emphasize that, 'A major finding of this study—that by every economic measure neonatal intensive care for infants weighing 1000–1499 g is superior to neonatal intensive care for infants weighing 500–999 g—is robust with respect to all sensitivity analyses investigated.' (p. 1335).

10. Did the presentation and discussion of study results include all issues of concern to users?

__x__ YES   ____ NO   ____ CAN'T TELL

The authors did not rely upon a single index on which to base their conclusions but rather calculated their results using three different economic evaluation techniques (CEA, CBA and CUA) each of which incorporates different value judgements. It is left to the reader to decide which index is most appropriate for evaluating NIC programmes. As well, the results are calculated for two distinct weight groups and three time horizons, again allowing the reader to judge the programme according to the criteria he or she feels are most important and relevant.

The authors explicitly acknowledge that many of the intangible factors that affect the value of NIC programmes have been excluded from the analysis (p. 1335, col. 1). It is unrealistic to expect, though, that these factors could be sufficiently quantified so as to permit inclusion. The authors' discussion of this point is a good demonstration of the principle that although certain costs and consequences may be immeasurable (for practical purposes), none the less they may be included in the analysis in a qualitative manner.

A great deal of discussion is devoted to the generalizability of the study results. A strong case is presented to support the assumption that the findings about health outcomes can be generalized to NIC programmes in similar urban settings during the period studied. With respect to the economic outcomes, costs and earnings appear to be generalizable over time and financing mechanisms, but perhaps not across settings where service intensities may vary.

Finally, the authors caution against comparing their results with the results of other studies (evaluating different interventions) in an attempt to assess the relative efficiency (or merit!) of NIC programmes. Differences in the viewpoint for the analysis, methodological approach or assumptions used to derive estimates of costs and outcomes may have dramatic effects on the results (as we saw earlier with the case of the discount rate). The authors are reluctant therefore to interpret their results normatively, and instead emphasize that, 'A judgement concerning the relative economic value of neonatal intensive care of very-low-birth-weight infants will require the economic evaluation of other health programmes by similar methods.' (p. 1336).

REFERENCES

Boyd, N. F., Sutherland, H. J., Ciampi, A., Tibshirani, R., Till, J. E., and Harwood, A. (1982). A comparison of Methods of Assessing Voice Quality in Laryngeal Cancer. In *Choices in health care: decision making and evaluation of effectiveness* (ed. R. B. Deber and G. G. Thompson), pp. 141–4. Department of Health Administration, University of Toronto.

Boyle, M. H., Torrance, G. W., Sinclair, J. C., and Horwood, S. P. (1983). Economic evaluation of neonatal intensive care of very-low-birth-weight infants. *N. Engl. J. Med.* **308**, 1330–7.

Cadman, D. and Goldsmith, C. H. (1982). Decision Making in Developmental Medicine. Presented at the Annual Meeting of the American Academy of Cerebral Palsy and Developmental Medicine, San Diego, October 6–10.

Churchill, D. N., Lemon, B. C., and Torrance, G. W. (1984a). A cost-effectiveness analysis of continuous ambulatory peritoneal dialysis and hospital hemodialysis. *Medical Decision Making* **4**, (4), 489–500.

—— Morgan, J., and Torrance, G. W. (1984b). Quality of life in end-stage renal disease. *Peritoneal Dialysis Bulletin* **4**, 20–3.

Hershey, J. C., Kunreuther, H. C., and Shoemaker, P. J. H. (1982). Sources of bias in assessment procedures for utility functions. *Management Science* **28**, (2), 936–54.

Holloway, C. A. (1979). *Decision making under uncertainty: models and choices*. Prentice-Hall, Englewood Cliffs.

Kahneman, D. and Tversky, A. (1982). The psychology of preferences. *Sci. Am. J.* Jan **1**, 160–73.

Kaplan, R. M. and Bush, J. W. (1982). Health related quality of life measurement for evaluation research and policy analysis. *Health Psychol.* **1**, 61–80.

—— —— and Berry, C. C. (1976). Health status: types of validity of the index of well-being. *Health Services Research* **11**(4), 478–507.

Llewellyn-Thomas, H., Sutherland, H. J., Tibshirani, R., Ciampi, A., Till, J. E., and Boyd, N. F. (1982). The measurement of patients' values in medicine. *Medical Decision Making* **2**, 449–62.

Logan, A. G., Milne, B. J., Achber, C., Campbell, W. P., and Haynes, R. B. (1981). Cost-effectiveness of worksite hypertension programme. *Hypertension* **3**, (2), 211–18.

McNeil, B. J., Weichselbaum, R., and Pauker, S. G. (1981). Speech and survival: tradeoffs between quality and quantity of life in laryngeal cancer. *N. Engl. J. Med.* **305**, (17), 982–7.

von Neumann, J. and Morgenstern, O. (1953). *Theory of games and economic behaviour* (3rd edn). Wiley, New York.

Pliskin, J. S., Shepard, D. S., and Weinstein, M. C. (1980). Utility functions for life years and health status. *Operations Research* **28**, (1), 206–24.

Sackett, D. L. and Torrance, G. W. (1978). The utility of different health states as perceived by the general public. *J. Chronic Dis.* **31**, (11), 697–704.

Sutherland, H. J., Dunn, V., and Boyd, N. F. (1983). Measurement of values for states of health with linear analog scales. *Medical Decision Making* **3**, (4), 477–87.

Torrance, G. W. (1976). Social preferences for health states: An empirical evaluation of three measurement techniques. *Socio-Economic Planning Sciences* **10**, (3), 128–36.

—— (1982). Preferences for health states: a review of measurement methods. In *Clinical and economic evaluations of perinatal programmes* (ed. J. C. Sinclair), Mead Johnson Symposium on Perinatal and Developmental Medicine No. 2, Vail, Colorado.

—— (1984). Health states worse than death. In *Third international conference on systems science in health care* (ed. W. von Eimeren, R. Engelbrecht, and C. D. Flagle). Springer Verlag, Berlin.

——, Boyle, M. H., and Horwood, S. P. (1982). Application of multi-attribute utility theory to measure social preferences for health states. *Operations Research* **30**, (6), 1043–69.

——, Sackett, D. L., and Thomas, W. H. (1973). Utility maximization model for programme evaluation: a demonstration application. In *Health status indexes* (ed. R. L. Berg), pp. 156–71. Hospital Research and Educational Trust, Chicago.

——, Thomas, W. H., and Sackett, D. L. (1972). A utility maximization model for evaluation of health care programmes. *Health Services Research* **7**, (2), 118–33.

Weinstein, M. C. (1981). Economic assessments of medical practices and technologies. *Medical Decision Making* **1**, (4), 309–30.

—— and Stason, W. B. (1977). Foundations of cost-effectiveness analysis for health and medical practices. *N. Engl. J. Med.* **296**, (13), 716–21.

Wolfson, A. D., Sinclair, A. J., Bombardier, C., and McGeer, A. (1982). Preference measurement for functional status in stroke patients: inter-rater and inter-technique comparisons. In *Values and long-term care* (ed. R. Kane and R. Kane), pp. 191–214. Lexington Books Div DC Health Publishers, Lexington MA.

7. Cost–benefit analysis

7.1. SOME BASICS

7.1.1. Introduction

In this chapter we discuss some of the additional issues that might confront someone undertaking a cost–benefit analysis (CBA). As was mentioned earlier, cost–benefit analysis is potentially a broad form of economic evaluation, although in practice the range of costs and benefits investigated is often restricted by measurement difficulties. For example, CBAs often restrict themselves to the narrowly defined economic changes brought about by treatments and programmes, such as changes in health care costs and the productive output of patients (or those caring for them). The methodological issues raised by consideration of these items have been discussed in earlier chapters. This chapter also discusses the methods of valuing some of the more intangible costs and benefits, and the conceptual and practical issues that such valuation raises.

In addition, the chapter includes an exercise on valuation of intangible costs and benefits and a critical appraisal of a cost–benefit analysis of community-oriented versus hospital-based care for mental illness patients.

Finally, there is also a critical appraisal of a study which includes all of the forms of evaluation discussed in the book.

7.1.2. What are the different approaches to valuation of costs and consequences?

There are a number of methods of valuation, each of which has its own rationale. These include:

1. Market valuations—taking actual valuations where these exist (e.g., for most resource items) or imputing valuations by reference to the market price of similar commodities (e.g., the value of house-wives' time could be imputed by reference to the wages paid to domestic staff).

2. Clients' willingness-to-pay estimates—assessed directly (by asking them) or indirectly (by observing their behaviour). (E.g., asking

people what they would pay for a quicker form of travel, or observing the trade-offs they make between expenditure and travel time savings.)

3. Policy-makers' views—either explicitly stated or implicit in their actions. (E.g., the decisions made about building safety regulations could be used to impute policy-makers' valuations of human life.)

4. Practitioners' views or professional opinions—such as those on the appropriateness of different forms of care for given categories of patients. (E.g., court awards might be used to impute a value for the unpleasantness of a disfiguring injury).

7.1.3 What are the principles underpinning the various approaches to valuation?

This book has had a practical focus, since it is written for those health care professionals wishing to evaluate alternative treatments or programmes from an economic perspective. However, it is important for those undertaking economic evaluations to have some insight into the underlying theoretical principles. Whilst a thorough discussion of these issues is beyond the scope of the book, a few points are made below. Those wanting to pursue their interests further should consult Sugden and Williams (1979) or Drummond (1981).

The traditional approach to cost–benefit analysis in health care is firmly grounded in welfare economics theory. A key component of welfare economics is the Paretian value judgement.* The essential features of the Paretian value judgement are that:

1. A global efficiency optimum is reached when resources cannot be reallocated to make one person better off without making at least one person worse off.

2. The existing distribution of income is 'accepted' or can be treated as a separate issue (i.e., issues of efficiency can be treated separately from those of distributive justice).

Furthermore, although the Paretian value judgement says nothing about whose values are to count, Mishan (1975) points out that, 'Economists are generally agreed—rather as a canon of faith, as a political tenet, or as an act of expediency—to accept the dictum that each person knows his own interest best.'

* Named after an Italian economist based at the University of Lausanne in the late 19th century.

The main arguments for applying the Paretian approach to CBA in health care are that:

1. It is consistent with welfare economics principles.
2. It leads economists to concentrate on efficiency questions, leaving politicians and others to concentrate on the equity questions. (Indeed it will throw these issues into sharp relief.)
3. It forces consideration, by decision-makers, of clients' values. (These are often neglected by policy-makers, planners and clinicians.)

The main arguments against applying the Paretian approach are that:

1. Governments have subsidized health care in many countries. This is an indication that willingness-to-pay, particularly since it is constrained by ability to pay, should not be the main criterion for allocating health care resources.*
2. The Paretian approach as applied in practice does not actually require that the gainers from a given policy compensate the losers. The real challenge is in converting potential Pareto improvements into actual Pareto improvements (through taxes, subsidies, etc.). That is, one should not attempt to view the efficiency and distribution issues separately.
3. Application of a pure Paretian approach may alienate decision-makers. More progress can be made by encouraging more systematic decision-making in health care, and by revealing implicit values.
4. It is not clear that individuals want to have their preferences sought for every single decision. It may be that a more sensible way of upholding consumer sovereignty is for them to delegate choices to informed decision-makers.

The other approaches considered in the book [e.g., cost-effectiveness analysis (CEA) and cost–utility analysis (CUA)] are based on Paretian principles to the extent that the prices used on the resource side of the equation are assumed to have welfare significance. That is, the opportunity cost of the resources consumed can be viewed as the willingness-to-pay for the forgone project. Also, in CUA clients' valuations are sometimes used.

---

* Of course, in assessing individuals' willingness-to-pay for the consequences of health care programmes, one might be able to design hypothetical experimental situations where the respondents' actual ability to pay does not influence their valuations.

Therefore, viewed in this light, approaches such as CEA and CUA are partly based on Paretian principles. Conversely one could argue that, in any case, the prices for resources that one observes in the health care system do not reflect the social opportunity costs of those resources, as there is not a 'perfect market' for inputs such as doctors' time, etc. Therefore, economic evaluation is just a method of identifying relevant factors in making choices in health care and not directly related to identifying improvements in economic welfare.

Either way there does not appear to be total agreement on the appropriate conceptual formulation or practical application of economic evaluation in health care. Some authors (Sugden and Williams, 1979; Drummond, 1981) have argued for a *decision-making* approach to CBA. In this approach, the main contribution of CBA is that of encouraging a systematic approach to decision-making and to make values explicit, whatever their source. It is argued that this would be a significant improvement on existing decision-making procedures.

7.1.4. How far should valuation be pursued?

A common issue confronting evaluators undertaking a cost–benefit analysis is that of how far to go in valuing some of the more intangible items. Two criteria for coming to a view on this issue have already been introduced, namely:

1. Is it likely that the gathering of more information on the intangible items will change the result of the study?
2. Are the costs of gathering the information affordable?

Here a third criterion can be added: Will valuation of the intangible outcomes encourage more informed decision-making or not? That is, although it could be argued that attaching money values to the intangible items leads to more explicit consideration of them, it could be argued that valuation is an excuse for less thoughtfulness on the part of decision-makers. They may take the estimates at face value and not question the method of valuation. They may unknowingly adopt values to which they do not subscribe. Therefore, the presentation of the intangible costs and benefits in the analysis is always a matter for careful consideration. Very little is known about decision-makers' perceptions of mixtures of quantified and unquantified information.

7.1.5. Should issues of distributive justice be tackled?

Economic evaluation is concerned with the economic efficiency of alternative treatments or programmes. However, there are other important considerations in making choices in the health care field—one such consideration being equity and distributive justice. Some economists have attempted to incorporate this dimension into their analysis, by weighting costs and consequences depending on the persons upon which they fall. (Usually the weights relate to the income or wealth of individuals. A dollar to a poor person is weighted more heavily than a dollar to a rich person.)

There is no clear view on this issue in the literature. The comments we would make are that:

(a) one should note these other considerations in the discussion of study results; and

(b) the absence of explicit weights does not mean that costs and benefits in the analysis are in any sense 'unweighted'. If costs and benefits are summed, or combined in a cost–benefit index, they are implicitly weighted equally. Also, as mentioned in Section 7.1.3, no estimations are 'value free', or independent of the existing distribution of income or existing values and ideologies.

7.2. VALUING 'INTANGIBLE', OR UNPRICED, ITEMS IN COST-BENEFIT ANALYSIS: EXERCISE

7.2.1. Description of the situation

Suppose that the National Cancer Institute is planning a cost–benefit study of whether it is worthwhile instituting breast cancer screening (by mammography) for women aged 40 to 59 years old and that you have been commissioned to undertake the analysis.

Assume for the moment that the viewpoint for the analysis is to be that of society and that most of the cost items, such as the medical resources required to mount the screening programme and the costs of hospitalization for cancer victims, can be estimated fairly unambiguously.

7.2.2. Tasks

What approach would you adopt to the valuation of the following more 'intangible' items?

1. Womens' time taken up in obtaining the screening test (e.g., work time, leisure time).
2. The 'psychic' element of treatment costs (which may be averted by early detection) (e.g., the side effects of chemotherapy for more advanced stages of the disease).

3. The value to women of being reassured that they do not have breast cancer.
4. The value of life-years saved by successful early detection and treatment of the disease.
5. The time given by relatives in home nursing of women suffering from the disease.

7.2.3. Solution

(A) The first major issue in deciding upon one's approach to the valuation of 'intangible' items is whether one believes it would be useful to attempt this at all. ('Useful' in this case relates to whether valuation of these items would lead to a more informed decision.)

As an alternative to valuation of the health effects in dollar terms, one might adopt the cost-effectiveness or cost–utility approach, expressing the economic results in terms of cost per life-year or cost per quality-adjusted life-year gained.

(B) The second major issue to be resolved is whether one adopts an approach consistent with Paretian principles, or whether one is willing to consider a more broadly based approach to valuation—perhaps presenting a range of estimates derived by different methods.

(C) Some possible approaches to the valuation of the items identified are:
   1. *Women's time taken up in obtaining the screening test.*
      (a) lost earnings (for women in employment):
         — taken gross (of taxes and employee benefits) these are an estimate of how time is valued in a productivity capacity;*
         — taken net, these are a measure of what the sacrifice in time is worth to the women (see (3) below);
      (b) imputed earnings (for women not in the labour force). This approach embodies the same logic as the gross earnings approach above (e.g., what would it cost to replace housewives' services?);
      (c) time values, as estimated from modal choice analyses (i.e., observations of choice of mode of travel).

---

* See the comments in Chapters 3 and 5 concerning the problems in taking wage losses as an estimate of the value of lost production.

2. *'Psychic' element of treatment costs (e.g., side effects of chemo-therapy)*
   (a) expenditure by patients on measures to reduce this kind of unpleasantness (e.g., on medication);
   (b) direct measurement by questionnaire or interview.
   In general this is a difficult item; it may be better approached through utility measurement.

3. *Value of being reassured*
   (a) direct measurement by questionnaire or interview;
   (b) observation of how much women are willing-to-pay in order to obtain reassurance (e.g., sacrificing time and expense to attend for screening).

4. *Value of life-years saved*

   A number of approaches could be used. For example:
   (a) direct measurement; willingness-to-pay for changes in risk (e.g., the approach advocated by Jones-Lee, 1976);
   (b) inference from behaviour (e.g., attitudes to personal safety);
   (c) gross discounted earnings (the 'human capital' approach);
   (d) implicit values from public policies (e.g., decisions about safety measures);
   (e) court awards (although often based on earnings);
   (f) explicit values for life used in public sector investments (e.g., such an approach is used in road transport appraisal in the UK. See Mooney, 1977, for more details);
   (g) individuals' expenditure on insurance (although this is more closely related to the value of one's life to others than the value of life to oneself).

5. *Time given by relatives in home nursing*

   It might be worth considering whether this is a cost, a benefit, or both. (This depends on whether the individual is at equilibrium in how he or she spends this time.) Possible approaches are:
   (a) lost income or other sacrifices in money or time;
   (b) the cost of replacing informal nursing help.

(D) Summary points
   1. The valuation of intangible items poses many problems.
   2. No method is free from conceptual or practical difficulties.
   3. Beware of unthinking application of the human capital approach (that is, valuing human life as the present value of discounted earnings streams).

4. Very often, valuation of all items in dollar terms may not be necessary or helpful.
5. Difficulties with valuation of 'intangibles' in dollar terms has steered many analysts towards cost–utility analysis.

7.3 CRITICAL APPRAISAL OF A PUBLISHED ARTICLE

Reference: Weisbrod, B. A., Test, M. A., and Stein, L. I. (1980). Alternative to mental hospital treatment. II Economic benefit—cost analysis. *Arch. Gen. Psychiatry* **37**, 400–5.

This article is assessed below using the 10 questions set out in Annex 3.1. It is suggested that you locate the article and attempt the exercise before reading the assessment given below.

1. Was a well-defined question posed in answerable form?

__x__ YES _____ NO _____ CAN'T TELL

The main question addressed by the study is, 'What are the comparative costs and benefits of a traditional hospital-based treatment approach and an experimental community-based alternative to mental hospital treatment?' A societal point of view is adopted requiring that all resource costs and consequences be considered regardless of to whom they accrue. However, given the detailed nature of the cost and benefit components presented in Table I of the article, the evaluation could easily be conducted from a variety of viewpoints.

2. Was a comprehensive description of the competing alternatives given?

_____ YES _____ NO __x__ CAN'T TELL

The reader is referred to an accompanying article in the same issue of the journal for more comprehensive descriptions of the competing alternatives (Stein, L. I. and Test, M. A. 1980. Alternative to mental hospital treatment: I Conceptual model, treatment programme and clinical evaluation. *Arch. Gen. Psychiatry* **37**, 392–7). Since the descriptions of the alternatives appear in the same journal issue as the present paper, this form of referencing is acceptable. However, had the background papers appeared in other issues, or other journals, a more elaborate summary would have been required.

A *do-nothing* alternative is not considered, as the question of interest is not whether treatment *per se* is worthwhile, but rather which mode of treatment produces a smaller increment of costs over benefits. (It is interesting to note that neither alternative 'pays for itself', at least not for the first 12 months of a patient's treatment.)

3. Was there evidence that the programmes' effectiveness had been established?

_____ YES _____ NO __x__ CAN'T TELL

This economic evaluation was undertaken as part of a randomized controlled trial conducted over a three-year period. Few details of the relative effectiveness of the two alternatives are provided and the reader is again referred to the accompanying paper cited above. Of course the mere existence of the paper does not imply that effectiveness has been established; the quality of the evidence can only be determined by consulting the suppporting reference.

4. Were all the important and relevant costs and consequences for each alternative identified?

__x__ YES _____ NO _____ CAN'T TELL

The evaluation is conducted from a societal viewpoint and consequently all possible costs and benefits must be considered. The authors have certainly compiled an impressive list of the various costs and benefits associated with each alternative (Table I).

The costs considered include the direct treatment costs as well as the 'indirect' treatment costs to a number of public agencies other than the providing institution. Note that the authors do not define the term 'indirect costs' in the generally accepted manner (i.e., the production losses or costs of time lost from work for patients and their families). Rather, they use this term to signify all treatment costs not falling directly on either the Mendota Mental Health Institute or the Experimental Centre Programme. Other costs considered are maintenance costs of an individual, family burden costs, and costs borne externally to the health sector, patients, and their families (e.g., law enforcement costs and illegal activity costs).

The benefits identified are mainly labour market benefits (i.e., indirect benefits), but items such as labour market behaviour and improved consumer decision-making are considered as measures of the programmes' relative success. Differences in clinical symptomatology and patient satisfaction with life are also identified as indicative of the benefits associated with improved patient mental health.

5. Were costs and consequences measured accurately in appropriate physical units?

___x___ YES _____ NO _____ CAN'T TELL

Not all the costs and benefits identified for inclusion were able to be measured using monetary values, thus for those items for which the authors were unable to develop monetary estimates, they calculated quantitative nonmonetary , indicators. Although these nonmonetary estimates could not be incorporated directly into the benefit–cost calculations, quantifying them in nonmonetary terms proved useful; it appears that some of the intuitive predictions about the relative impact of each programme on certain costs and benefits were demonstrated to be unfounded. For example, the authors initially suspected that, 'Because the experimental (E) group spent less time in the mental hospital and also less time in other hospitals, we expected they might have more frequent and serious encounters with the law.' But the figures indicate that the average number of arrests is actually lower for the E group, suggesting greater social cost savings and possibly greater socially accepted behaviour for the group.

Joint costs of the hospital plant and land were apportioned accordingly, although there is no mention of the technique employed. The costs of the E programme which were attributable to research rather than treatment were appropriately deducted.

An interesting issue arises in the actual measurement of the E programme costs. The authors state that costs were allocated '... in a manner that was designed to determine what average cost would have been if the centre had operated for the entire period in the way it operated at the middle of the programme period'. In other words, the average cost figures used to proxy resource use likely reflect the most efficient provision of the E programme. (The authors do note, however, that there is no way of determining whether the maximum patient population they observed is sufficient just to exhaust any potential economies of scale, or whether expanding the programme still further would induce yet more decrease in average cost.) It is not possible to tell, however, if the current provision of the traditional hospital-based programme represents the most efficient use of hospital resources. Thus it might be possible that the study is comparing two programmes, one of which is operating at optimal output and one of which is not. If this were true, then the results are unfairly biased toward the programme which is operating at minimum average cost—the E programme.

Benefits are measured as the earnings from competitive employment and from sheltered workshops. The authors recognize that this approach to measuring indirect benefits neglects the possibility that better health is desirable in its own right, apart from its contribution to labour productivity or earnings.

6. Were costs and consequences valued credibly?

___x___ YES   _____ NO   _____ CAN'T TELL

Costs for category I items (see Chapter 3, Fig. 3.1) are estimated using market prices. Although the authors derive a *per diem* hospital cost, the figure is not the typical *per diem* found in hospital financial accounts. Their figure is calculated by considering both operating and capital costs, converting these to a daily figure, and multiplying this by the average number of days of hospitalization for each group. Where market prices were not available (e.g., administrative services that were provided to the E programme without charge) imputed values were estimated to reflect true resource use.

Data for cost categories II and III were gathered from patients and family interviews as well as from contact with various social service and community agencies. Maintenance costs were derived primarily from the agencies involved, as well as from patients themselves.

The logic for including maintenance costs is that hospital cost estimates include allowances for food and shelter, etc., and thus for consistency these costs should be calculated for the E programme as well. But there may be some doubt about the use of welfare payments to value the maintenance costs of patients; a preferable method would be to rely on the actual expenditures by patients and families on maintenance items.

Valuation of labour market benefits was determined using earnings data obtained from patients. The implication of this approach is that prevailing wage rates accurately reflect true productivity. But the authors note that earnings estimates were obtained during a period of recession; thus it is possible that observed wages represent an underestimate of the productive contribution of labourers.

Expenditure on insurance and data on the proportion of patients having a savings account were used as indicators of the degree of programme success. However, both these items are partly a function of total earnings and may consequently act as proxies for income. If so, then double-counting of benefits would occur if both success indicators and earnings were used simultaneously as measures of benefits.

7. Were costs and consequences adjusted for differential timing?

  x   YES   _____ NO   _____ CAN'T TELL

Discounting is not employed, since costs and benefits are estimated for a one year programme only. However the capital costs of the institutional option are annuitized at 9 per cent, a rate chosen to reflect the rate of return on the market value of the capital costs.

8. Was an incremental analysis of costs and consequences of alternatives performed?

  x   YES   _____ NO   _____ CAN'T TELL

The incremental benefits of the E programme over the C (control) programme, compared with the incremental costs it generates, were calculated using the following net benefit formula:

$$(B_E - B_C) - (C_E - C_C) = (\$2364 - \$1168) - (\$8093 - \$7296) = \$399$$

The net difference between benefits and costs of the two programmes is $399, or approximately 5 per cent of the cost of the experimental programme.

9. Was a sensitivity analysis performed?

_____ YES   x   NO   _____ CAN'T TELL

No extensive sensitivity analysis is employed, although there is a good discussion of potential sources of variations in costs. Although a formal sensitivity analysis is an important element of a sound economic evaluation, the omission in this study is not as severe as it might have been, had the cost and benefit estimates not been calculated with such precision and attention to detail.

10. Did the presentation and discussion of study results include all issues of concern to users?

  x   YES   _____ NO   _____ CAN'T TELL

The analysis concludes that, considering all the forms of benefits and costs that the authors were able to derive in monetary terms, the E programme provides both additional benefits and additional costs when compared with the conventional treatment approach, but the added benefits are nearly $400 more per patient per year than the added costs.

Although these results provide a useful insight into the differential costs and benefits incurred over a one-year period, the authors acknowledge that the real issue of interest is the present discounted value of the lifetime benefits and costs of alternative treatment modes. Presumably, if the experimental programme is adopted on an ongoing basis, benefits and costs would continue, but not necessarily at the levels found for the first year. This is especially true for the benefits resulting from increased earnings; the benefits accruing to a single individual would not be merely the addition of one year's income, but rather the present value of the stream of all future earnings.

The authors provide a strong warning about generalizing the study results to other settings. They suggest a number of factors which may affect the cost and benefit assessments, some of which include: the level or mix of treatment resources used, the scale of provision of the services, the size of the community in which the programmes are offered and the type of community-based programmes recommended.

An important conclusion emerges with respect to the viewpoint employed. Although the experimental progamme is preferable from the societal point of view, it is actually more costly to the health sector and to some public agencies. Readjustments in budgets would therefore be necessary if the experiment was to be extended.

As well, the study admittedly neglects the potential for adverse public reaction to the community-based programme. Extra resources may be required, say, in public relations activities, to alleviate the community's concerns.

Finally, one question not addressed by the paper is that of the balance of care between hospital and community, based on the patient degree of dependency (i.e., which patients are best cared for in each location), although in fairness to the authors this was not an intended objective of the analysis.

7.4. CRITICAL APPRAISAL OF A PUBLISHED ARTICLE

Reference: Stason, W. B. and Weinstein, M. C. (1977). Allocation of resources to manage hypertension. *N. Engl. J. Med.* **296**, 732–9.

Cost–benefit analysis

As a final critical appraisal exercise, you are asked to consider the article by Stason and Weinstein (1977). Although primarily a cost-effectiveness analysis, it embodies a number of the methdological approaches described in this book. The article is assessed below using the 10 questions set out in Annex 3.1. It is suggested that you locate the article and attempt the exercise before reading the assessment given below.

1. Was a well-defined question posed in answerable form?

___x___ YES   _____ NO   _____ CAN'T TELL

First, the study is concerned both with the resources, or inputs, going into the detection/treatment of hypertension and the effects, or outputs, of detection/treatment in terms of increased years of life expectancy adjusted for changes in the quality of life. (Note that the authors do not distinguish between 'effects' and 'benefits' in the usual way.) Second, on p. 732 of the paper they specify four questions for analysis, all of which involve, explicitly or implicitly, comparison of alternatives. Their first question compares treatment of hypertension to 'doing nothing'; that is, treatment of the cardiovascular morbidity resulting from hypertension. The second question, 'How efficient a use of resources is the treatment of essential hypertension?' implicitly compares hypertension treatment to 'other uses of health care resources', as we find out later in the article (p. 738, col. 1). Naturally, even with data on the cost-effectiveness of treating hypertension, the authors find their second question difficult to answer. The third question again involves implicit comparisons, this time among programmes aimed at different age, sex and pretreatment diastolic blood pressure groups. The fourth question explicitly states a comparison between screening programmes to detect hypertension and improved efforts to manage known hypertensives. Finally, a societal viewpoint is specified (p. 732, col. 1). An interesting decision context is also specified, primarily in conjunction with the fourth question. The authors ask how a $1 million budget should be allocated so as to achieve maximum 'health benefit' from a hypertension programme.

2. Was a comprehensive description of the competing alternatives given?

_____ YES   ___x___ NO   _____ CAN'T TELL

This assessment is debatable given that the authors are dealing with a macro analysis rather than a specific programme operating out of a specific facility in a known location, and given that the authors refer readers to their detailed book on the same subject. Nevertheless, there are at least four activities or programmes involved in the analysis (1) screening, (2) treatment of hypertension, (3) treatment of cardio-vascular morbidity in untreated hypertensives, i.e., the *do-nothing* option, and (4) adherence intervention package. The alternatives usually involve some combination of these, yet it is not easy to tell who does what to whom, where and how often in any of the four activities. On p. 734 the authors present a multistage model of the process of managing hyper-tensives, which implies a detailed knowledge of all activities, yet frequently only the assumed costs for the activities are provided. The description of the activities which underpins the cost calculation does not generally appear.

3. Was there evidence that the programmes' effectiveness had been established?

__x__ YES   _____ NO   _____ CAN'T TELL

The authors cite two studies, references 1 and 5, in asserting that treat-ment of hypertension is effective at least in selected patients (p. 732, col. 1) and that hypertension control yields mortality and morbidity 'benefits' (p. 732, col. 2). Although it is not possible to tell from this article whether the studies referred to were randomized controlled trials, the authors seem justified in relying on the general knowledge of their audience that hypertension is one of the few conditions where effective-ness of treatment has been established quite rigorously. The authors further allow for variations in the 'benefit' (i.e., reduced risk of strokes and myocardial infarctions from untreated hypertension) of hyper-tension treatment by using three assumptions: full-benefit, half-benefit, and age-varying partial-benefit (p. 732, col. 2). The effectiveness of the screening programme and especially the adherence intervention package are dealt with by simulating different performance levels.

4. Were all the important and relevant costs and consequences for each alternative identified?

_____ YES   _____ NO   __x__ CAN'T TELL

The formulation of the cost-effectiveness ratio (i.e., what is included and what is excluded) presented here is controversial! They include in costs the cost of antihypertensive treatment and the cost of treating side effects of antihypertensive treatment, both fairly standard operating costs of the 'programme'. [Later in the article, in their exercise on allocating a fixed budget (p. 734, col. 1) they expand these programme costs to include the screening and adherence intervention packages.] Although some information is given on the composition of these programme costs (e.g., physician visits, laboratory examinations, antihypertensive medications), it is not completely clear what components have been included in the charges and cost assumptions.

More controversial is their inclusion, in costs, of the cost of treating noncardiovascular illness in added years of life. Some would argue that this burdens the 'programme' unfairly with costs extraneous to it; others would contend that if one is conducting a planning exercise for a system, the relevant question is what happens to system costs if a new programme goes ahead.

Note that they also subtract from programme costs the expected savings in treatment costs associated with reductions in the frequency of cardiovascular morbid events, thus arriving at 'net expected medical care costs'. These expected savings are sometimes treated as a direct benefit of the programme, and could be placed in the denominator of a cost–benefit ratio.

The authors say that their analytic viewpoint is that of society (p. 732, col. 1); however, it appears that the viewpoint is a combination of the health care sector and a partial patient viewpoint. If it were truly a societal viewpoint, we might expect to see increased work time (sometimes called production gains, or indirect benefits) valued somehow. Similarly, any production losses or indirect costs associated with patients receiving the programme would have been included. (This might not be as minor as it sounds since there is apparently a 'labelling' phenomenon which occurs with hypertensives that leads to increased absenteeism from work once they assume the sick role.) Finally, even from a strictly patient viewpoint we might have expected to see additional out-of-pocket expenses (though likely minor) other than charges for health care services included.

Effects clearly incorporate the patient viewpoint since they are measured as the increase in life expectancy, adjusted for the quality of life without nonfatal cardiovascular events, but with the side effects of medication.

5. Were costs and consequences measured accurately in appropriate physical units?

_____ YES   _____ NO   _x_ CAN'T TELL

This may be a harsh assessment in a case like this, where a macro analysis, from a larger book, has been condensed into a journal article; however, one point to note is that it is generally not possible to see the list of 'ingredients' and their amounts, which are used up in the hypertension programme, or in the treatment of subsequent events in the 'no-programme' case. Some information is provided on antihypertensive treatment cost (three physician visits per year, p. 732, col. 2) in the basic cost-effectiveness analysis. Some information on quantities of resources is also provided when describing the multistage model ('Stage 6 involves a diagnostic evaluation at a cost of $100 followed by six months of treatment of $200 per year ...' p. 734, col. 1). But generally only the calculated or assumed costs are given, often as totals, rather than the individual quantities and prices of component costs. The measurement of effectiveness in life-years gained appears straightforward.

6. Were costs and consequences valued credibly?

__x__ YES   _____ NO   __x__ CAN'T TELL
(for costs)   (for consequences)

Costs are valued in market prices, based on charges for specific health care services and average expenditures for treatment of cardiovascular and non-cardiovascular events. There do not appear to be any special circumstances requiring adjustment of prices or costs. (A charge does not always reflect a resource cost, but there is no reason to assume that a better measure could be found, certainly not easily, nor that it would markedly affect this analysis.)

The valuation of quality of life is less satisfying, however (thus the 'can't tell' above). The authors state, 'It is assumed that the average patient would be willing to give up 3.5 days of life per year (i.e., 0.01 times life years) to be free of the expected subjective side effects of therapy, so that one year with the side effects was taken to be equivalent to 0.99 quality-adjusted life-years.' (p. 733, col. 1). It is not possible to tell whose values these 3.5 days represent. Are these the values of the authors, a random sample of physicians or of patients? Nor is it possible to tell how they were obtained—casually, or by some preference measurement technique. (Reading of the book suggests that they are primarily based on subjective estimates, made without preference measurement, by a group of physicians.) It does, however, appear that utility analysis is appropriate in this case, since side-effects of antihypertensive medication are not insignificant and nonfatal cardiovascular events may dramatically affect quality of life. (Given that the authors compare cost to quality-adjusted life-years, this aspect of the study could be called a cost–utility analysis. See Chapter 6 of this book for more details.)

7. Were costs and consequences adjusted for differential timing?

___x___ YES _____ NO _____ CAN'T TELL

Both costs and effects are discounted at 5 per cent. The discounting of effects is recognized by the authors as controversial, and a justification is provided. (Other justifications also exist.) Although no justification is given for the 5 per cent rate, readers are referred to a methodology article by one of the authors. Moreover, discount rates of 0 per cent and 10 per cent are also employed in a sensitivity analysis.

8. Was an incremental analysis of costs and consequences of alternatives performed?

___x___ YES _____ NO _____ CAN'T TELL

The cost-effectiveness formulation is explicitly set up in this fashion (p. 733, col. 1) for the question of treating hypertension versus treating the events asociated with untreated hypertension. This is carried through, though sometimes implicitly, in the rest of the paper.

9. Was a sensitivity analysis performed?

___x___ YES _____ NO _____ CAN'T TELL

The authors are quite careful in examining the sensitivity of their analysis to changes in the values of critical variables. They explicitly include the discount rate (0 per cent and 10 per cent in addition to 5 per cent), medical treatment costs (range $100 to $300), and quality of life adjustment (2 per cent as well as 1 per cent of life expectancy) in a sensitivity analysis section (p. 733, col. 2). In addition, they state, rather vaguely, that they made 'wide variations in the imputed costs and quality of life adjustments for strokes and myocardial infarction'.

The results themselves are also reported in a way that gives a similar effect to that of sensitivity analysis. They examine the significance of variations in the level of diastolic blood pressure achieved and compliance with therapy. Alternative assumptions are also employed for variables (in the glossary on p. 732, col. 2).

Justifications are not provided in the article for the values used in the sensitivity analysis. Finally, note how the cost-effectiveness ratio is sensitive to the quality of life adjustment, the discount rate and the cost of hypertension treatment.

10. Did the presentation and discussion of study results include all issues of concern to users?

  x   YES   NO   CAN'T TELL

The authors provide quite a detailed presentation and discussion of study results. Their main contribution is to show the cost-effectiveness of hypertension treatment, in dollars per quality-adjusted life-year, under various assumptions about the values of critical variables. They acknowledge that a cost-effectiveness ratio cannot tell us, by itself, whether or not to institute the programme. It must be judged against a decision criterion (p. 735, col. 2; p. 738, col. 1) of what society (for example) deems an appropriate extra amount to pay for an extra life-year. The authors provide answers, albeit qualified ones, to their four original questions, and acknowledge the imperfect quality of their data. They briefly mention the possibility that patients (especially uninsured ones) may value future statistical health benefits differently from physicians. They also point out the implications of their results for future efficacy and effectiveness research. One item that they do not discuss, but which is addressed by Fein in an editorial later in the same issue of the journal [*N. Engl. J. Med.* (1977) **296**, 751] is that of equity. Screening, though apparently less cost-effective than management, may be desirable because it reaches those individuals that the private sector would neglect. (Fein more generally provides an excellent caveat, reminding us to remember not only what is included, but also what is omitted; not only what is measured, but what is not.)

REFERENCES

Drummond, M. F. (1981). Welfare economics and cost benefit analysis in health care. *Scottish Journal of Political Economy* **28**, 125–45.

Fein, R. (1977). Editorial. *N. Engl. J. Med.* **296**, 751.

Jones-Lee, M. W. (1976). *The value of life*. Martin Robertson, London.

Mishan, E. J. (1975). *Cost benefit analysis*. George Allen and Unwin, London.

Mooney, G. H. (1977). *The valuation of human life*. MacMillan, London.

Sugden, R. and Williams, A. (1979). *The principles of practical cost–benefit analysis*. Oxford University Press.

8. How to take matters further

8.1. ECONOMIC EVALUATOR'S SURVIVAL GUIDE

8.1.1. Introduction

Like most other resources, the resources required to undertake economic evaluations are scarce. Therefore it is important to ensure that these resources are used profitably. (From the individual evaluator's point of view it is also important to avoid wasting one's time when this could be better spent on other activities.)

In this short section we present a list of questions the economic evaluator should ask himself, or another party requesting an evaluation, when embarking on a new study. The object is to minimize the following two difficulties:

(a) evaluators becoming involved in inappropriate or unprofitable evaluations; and

(b) evaluators spending longer than necessary on any given evaluation.

There is no strong scientific basis for the suggested questions proposed in Section 8.1.2, although some of them mirror quite closely the checklist for critically assessing the literature presented in Annex 3.1. Rather our questions reflect years of experience in participating in economic evaluations and the mistakes we have made. We can make no guarantees of course: disease and infirmity among the evaluation team, changes in government and world wars can disrupt the best laid plans! However, we feel that the book at least gives the evaluator a fighting chance.

Here's hoping that you survive and can do better than ourselves in the future!

8.1.2. Some questions to ask yourself when beginning a study

1. *Who needs this study and why?*

   (a) What viewpoints are legitimate/feasible?

(b) Is the person requesting the evaluation arguing for an unnecessarily restrictive viewpoint?

(c) Is anyone serious about the evaluation; that is, are the results likely to change any actions/policies that are being contemplated?

(d) Will it be possible for someone to act on the results of the evaluation, whatever these turn out to be; that is, are the necessary management procedures or decision-making procedures in place?

(e) In general do people have an open mind with respect to the evaluation results?

2. *How did we arrive at these alternatives for consideration?*

(a) Is more than one alternative programme proposed, or is the implicit comparison the status quo?

(b) What would happen if we did nothing at all?

(c) Have any important alternatives been omitted?

(d) Is this particular approach to meeting the given service objectives suggested by previous research, or does it represent someone's 'pet scheme'?

(e) Would slightly more or slightly less of the proposed programme be preferable; what would we lose if the progamme were pruned; what could be gained if extra features were added?

3. *What do we know about the effectiveness of the proposed alternatives*

(a) Have any of the alternative programmes, especially the one(s) now being proposed for economic evaluation, been shown to do more good than harm by controlled study (especially involving random allocation of subjects to programmes or treatments)?

(b) If so, what do we know about the methodological quality of that study and how do we know the same results would be obtained in our setting?

(c) If not, are the supporters of this economic evaluation serious about undertaking a controlled study of effectiveness of the new programme compared to existing approaches? (If not, are you serious about them?!)

(d) What justification can be given for going ahead with an economic evaluation without generating the effectiveness evidence?

4. *What do we know about the likely costs and funding implications of the proposed alternatives?*

(a) What would be your quick estimate (to the nearest $25 000) of the additional resources required to fund the new programme (if found to be effective)?

(b) Is this sum large in comparison to the likely costs (especially in your time input) of the evaluation? (If not, why are you involved at all?)

(c) Is there any hope that the extra resources will be found, either through cost savings generated by the programme, resources redeployed from elsewhere, or new funding?

(d) Would any such redeployment or new funding be hard to achieve?

5. *How would we carry out such an evaluation?*

(a) What resources would be needed for the evaluation (e.g. manpower, computing)?

(b) Are these already available or would extra support be required? (If so, has any thought been given as to where support might come from; in particular, are people aware of the methods of obtaining research grants?)

(c) What kind of moral support can you expect from those requesting the evaluation, especially within your own organization?

(d) When are the results of the evaluation required? Is everybody clear on what can be achieved (with the evaluation resources at your disposal) within the given time period?

(e) Whose cooperation do you need to undertake your study? (Will they give it willingly, or only if forced to?)

(f) Do you like the other people involved well enough to spend extended periods of time with them? (You may have to!)

8.2. ADDITIONAL LITERATURE

We cite a number of references below which relate to the forms of analysis described in this book.

1. *Cost analysis*

Horngren, C. T. (1982). *Cost accounting: a managerial emphasis* (5th edn). Prentice Hall, Englewood Cliffs, N. J.
[A good introductory text in cost accounting.]
Clements, R. M. (1974). *The Canadian hospital accounting manual supplement*. Livingston Printing, Toronto.
[Cost accounting for hospitals. Describes the step down and the step down with iterations methods for overhead allocation.]

Levin, H. M. (1975). Cost-effectiveness analysis in evaluation research. In *Handbook of evaluation research* (eds. M. Guttentag and E. Struening) Vol. 2, pp. 89–122. Sage Publishing, Beverly Hills.

[A good description of the ingredients approach to costing.]

Kaplan, R. S. (1973). Variable and self-service costs in reciprocal allocation models. *The Accounting Review* **XLVIII**, 738–48.

[Describes the theory for the simultaneous method of overhead allocation.]

Richardson, A. W. and Gafni, A. (1983). Treatment of capital costs in evaluating health care programmes. *Cost and Management* Nov–Dec: 26–30.

[A detailed description of the method for costing equipment and other capital items. Examples of studies that have done it incorrectly, and the consequences.]

Boyle, M. H., Torrance, G. W., Horwood, S. P., and Sinclair, J. C. (1982). A cost analysis of providing neonatal intensive care to 500–1499 gram birth weight infants. Research Report #51, Programme for Quantitative Studies in Economics and Population, McMaster University, Hamilton.

[A detailed example of costing episodes of hospital care. A comprehensive view of costs was taken. Overheads were allocated by the simultaneous method. Methods are applicable to any kind of institutional care—not just hospitals.]

2. *Cost-effectiveness analysis*

Methods

Drummond, M. F. (1980). *Principles of economic appraisal in health care*. Oxford University Press.

Warner, K. E. and Luce, B. R. (1982). *Cost–benefit and cost-effectiveness analysis in health care.* Health Administration Press, Ann Arbor.

[Both of the above books provide a comprehensive coverage of cost-benefit and cost-effectiveness analysis in health care evaluation.]

Weinstein, M. C. and Stason, W. B. (1977). Foundations of cost-effectiveness analysis for health and medical practices. *N. Engl. J. Med.* **296**(13), 716–21.

[This is a widely cited methodological reference that deals with several important issues (e.g., discounting effects, viewpoints, formulation).]

Shepard, D. S. and Thompson, M. (1979). First principles of cost-effectiveness analysis in health. *Public Health Reports* **94**(6), 535–43.

[This is a straightforward explanation of CEA, with illustrations; however, like Weinstein and Stason, these authors do not explicitly distinguish cost–utility from cost-effectiveness analysis.]

Levin, H. M. (1975). Cost-effectiveness analysis in evaluation research. In *Handbook of evaluation research* (eds M. Guttentag and E. L. Struening) Vol. 2, pp. 89–122. Sage Publications, London.
[Levin presents a relatively 'narrow' view of the components to be included in a CEA formulation, compared to other analysts. However, he also presents a very thorough treatment of costs.]

Examples

Stason, W. B. and Weinstein, M. C. (1977). Allocation of resources to manage hypertension. *N. Engl. J. Med.* **256**(13), 732–9.
[This is a companion piece to their methods article in the same issue. See also the cautioning note sounded by Rashi Fein in the critique of Stason and Weinstein in the same issue.]

Culyer, A. J. and Maynard, A. K. (1981). Cost-effectiveness of duodenal ulcer treatment. *Social Science and Medicine* **15C**, 3–11.
[An interesting and thoughtful treatment of several methodological issues in a situation where the clinical state-of-the art was still unsettled.]

Ludbrook, A. (1981). A cost-effectiveness analysis of the treatment of chronic renal failure. *Applied Economics* **13**, 337–50.
[Renal failure has been the subject of numerous econmic evaluations, referenced herein, and as such would provide good exposure to the ways in which different analysts approach the same problem with different methods.]

The following four papers contain more recent examples of the application of cost-effectiveness analysis

Levine, M. N., Drummond, M. F., and Labelle, R. J. (1985). Cost-effectiveness in the diagnosis and treatment of carcinoma of unknown primary origin. *Canadian Medical Association Journal* **133**, 977–87.
[A recent cost-effectiveness analysis employing a decision tree.]

Churchill, D. N., Lemon, B. C., and Torrance, G. W. (1984). A cost-effectiveness analysis of continuous ambulatory peritoneal dialysis and hospital hemodialysis. *Medical Decision Making* **4**(4), 489–500.
[A well done cost-effectiveness analysis comparing a new dialysis technology with an existing one.]

Fulton, M. J. and Barer, M. L. (1984). Screening for congenital dislocation of the hip: an economic appraisal. *Canadian Medical Association Journal* **130**, 1149–56.
[A particularly interesting example of sensitivity analysis.]

Berwick, D. M. and Komaroff, M. D. (1982). Cost-effectiveness of lead screening. *N. Engl. J. Med.* **306** (23), 1392–8.
[Another good example of economic evaluation appied to screening procedures.]

3. *Cost–utility analysis*

Utility theory

Raiffa H. (1968). *Decision analysis: introductory lectures on choices under uncertainty*. Addison-Wesley, Reading, Mass.
[Still one of the best introductory texts on utility theory.]
Keeney, R. L. and Raiffa, H. (1976). *Decisions with multiple objectives: preferences and value tradeoffs*. Wiley, New York.
[Comprehensive coverage of single-attribute and multi-attribute utility theory. Quite mathematical and heavy in parts.]
Torrance, G. W. (1976). Toward a utility theory foundation for health status index models. *Health Services Research* **11** (4), 349–69.
[Re-casts utility theory into the health domain. Relatively mathematical and heavy.]

Cost–utility analysis in health

Torrance, G. W., Thomas, W. H., and Sackett, D. L. (1972). A utility maximization model for evaluation of health care programmes. *Health Services Research* **7** (2), 118–33.
[The original cost–utility model as first developed.]
Bush, J. W., Chen, M. M., and Patrick, D. L. (1973). Health status index in cost-effectiveness: analysis of PKU programme. In *Health status indexes* (ed. R. L. Berg). Hospital Research and Educational Trust, Chicago.
[Applications to PKU screening. Utilities measured by rating scale. Direct costs only.]
Stason, W. B. and Weinstein, M. C. (1977). Allocation of resources to manage hypertension. *N. Engl. J. Med.* **296** (13), 732–9.
[Application to hypertension. Hypothetical utilities. Direct costs only.]
Torrance, G. W. and Zipursky, A. (1984). Cost-effectiveness of ante-partum prevention of Rh immunization. *Clinics in Perinatology* **11** (2), 267–81.
[A recent study which also gives a listing of cost-utility results from other studies.]

Weinstein, M. C., Pliskin, J. S., and Stason, W. B. (1977). Coronary artery bypass surgery: decision and policy analysis. In *Costs, risks, and benefits of surgery* (eds J. P. Bunker, *et al.*). Oxford University Press, New York.

[Application to coronary artery bypass surgery. Hypothetical utilities. Direct costs only.]

Weinstein, M. C. (1980). Estrogen use in postmenopausal women—costs, risks and benefits. *N. Engl. J. Med.* **303** (6), 308–16.

[Application to estrogen therapy. Hypothetical utilities. Direct costs only.]

Kaplan, R. M. and Bush, J. W. (1982). Health-related quality of life measurement for evaluation research and policy analysis. *Health Psychology* **I** (1), 61–80.

[Excellent review of the ten years of work by Bush, Kaplan, and colleagues at the University of California, San Diego.]

Williams, A. (1985). Economics of coronary artery bypass grafting. *British Medical Journal* **291**, 326–9.

[A thought-provoking paper providing a ranking of various procedures in terms of their cost per quality-adjusted life-year gained.]

4. *Cost–benefit analysis*

Theory

Mishan, E. J. (1975). *Cost-benefit analysis*. George Allen and Unwin, London.

Sugden, R. and Williams, A. H. (1979). *The principles of practical cost-benefit analysis*. Oxford University Press.

Dasgupta, A. K. and Pearce, D. W. (1972). *Cost–benefit analysis: theory and practice*. Macmillan, London.

[All three volumes provide a good introduction to the theory of CBA.]

Examples

Acton, J. P. (1975). *Measuring the social impact of heart and circulatory disease programmes: Preliminary framework and estimates*. Rand Corporation, R-1697-NHLI, Santa Monica.

[One of the first applications of willingness-to-pay measurements to health care evaluation.]

Drummond, M. F. (1981). Welfare economics and cost benefit analysis in health care. *Scottish Journal of Political Economy* **28** (2), 125–45.

[An examination of the relationship of theory to practice in CBA.]

Jones-Lee, M. W. (1976). *The value of life—an economic analysis*. Martin Robertson, London.

[A good theoretical treatment of the willingness-to-pay approach.]

Klarman, H. E. (1982). The road to cost-effectiveness analysis. *Milbank Memorial Fund Quarterly* **60** (4), 585–604.
[A good summary of the difficulties associated with CBA, by one of the first economists to apply CBA to health programmes.]
Mooney, G. H. (1977). *The valuation of human life.* MacMillan, London.
—— (1978). Human life and suffering. In *The valuation of social cost* (ed. D. W. Pearce). George Allen and Unwin, London.
[These two items give a good summary of the different approaches to valuing life.]
Schoenbaum, S. C., McNeil, B. J., and Kavet, J. (1976). The swine-influenza decision. *N. Engl. J. Med.* **295** (14), 759–65.
[Interesting use of the delphi technique to obtain secondary data quickly.]
Thompson, M. S., Read, J. L., and Laing, M. (1984). Feasibility of willingness-to-pay measurement in chronic arthritis. *Medical Decision Making* **4** (2), 195–215.
[Illustrates well the problems in obtaining accurate and meaningful willingness-to-pay measurements.]

5. *Further examples of empirical work in this field*

Many more examples of economic evaluations of health care programmes can be found in the following texts.

Drummond, M. F. (1981). *Studies in economic appraisal in health care.* Oxford University Press.
Drummond, M. F., Ludbrook, A., Lowson, K. V., and Steele, A. (1986). *Studies in economic appraisal in health care*, Vol. 2. Oxford University Press.
Warner, K. E. and Luce, B. R. (1982). *Cost–benefit and cost-effectiveness analysis in health care*. Health Administration Press, Ann Arbor.
[Together, these three texts review more than 200 studies.]

6. *Interesting discussions of the usefulness/limitations of economic evaluation*

Loewy, E. H. (1980). Cost should not be a factor in medical care. (Letter) *N. Engl. J. Med.* **302** (12), 697.
[And a thoughtful essay by Williams (see below) who argues the opposite.]
Williams, A. (1983). Medical ethics: health service efficiency and clinical freedom. *Nuffield/York Portfolios*, Folio 2, Nuffield Provincial Hospitals Trust, London.

Fuchs, V. R. (1980). What is CBA/CEA, and why are they doing this to us? *N. Engl. J. Med.* **303** (16), 937–8.

Fein, R. (1977). But on the other hand: high blood pressure, economics and equity, *N. Engl. J. Med.* **296** (13), 751–3.

Vladeck, B. C. (1984). The limits of cost-effectiveness. *American Journal of Public Health* **74** (7), 652–3.

Centerwall, B. S. (1981). Cost–benefit analysis and heart transplantation. *N. Engl. J. Med.* **304** (15), 901–3.

Author index

Author index

Horngren, C. T. 43, 46, 67, 170
Horwood, S. P. 14, 16, 27, 34, 42, 66,
    115, 119, 139, 147, 148, 171
Hull, R. D. 11, 17, 27, 29, 34, 35, 45,
    67, 94, 111
Hutton, R. C. 1, 4

Jones-Lee, M. W. 155, 167, 174

Kahneman, D. 116, 147
Kane, R. 148
Kaplan, R. M. 116, 117, 120, 147,
    174
Kaplan, R.S. 46, 67, 171
Kavet, J. 174
Keeler, E. 81, 111
Keeney, R. L. 173
Klarman, H. E. 174
Komaroff, M. D. 172
Knreuther, H. C. 116, 147

Labelle, R. J. 172
Laing, M. 174
Last, J. M. 17
Lemon, B. C. 116, 147, 172
Levin, H. M. 42, 67, 171, 172
Levine, M. N. 172
Lewicki, A. M. 7, 17, 30, 35, 43, 67
Llewellyn-Thomas, H. 119, 147
Loewy, E. H. 175
Logan, A. G. 11, 17, 100, 104, 111,
    114, 147
Lowson, K. V. 1, 4, 9, 17, 50, 61, 67,
    175
Luce, B. R. 1,4, 171, 175
Ludbrook, A. 1, 4, 11, 17, 172, 175

Maynard, A. K. 172
McGeer, H. 117, 148
McNeil, B. J. 116, 119, 148, 174
Milne, B. J. 11, 17, 100, 104, 111,
    114, 147
Milne, R. G. 13, 17
Mishan, E. J. 150, 167, 174
Mooney, G. H. 155, 167, 174
Morgan, J. 115, 147
Morgenstern, O. 118, 126, 148

Neuhauser, D. 7, 17, 30, 35, 43, 67
Neumann, J. von 118, 126, 148
Newell, D. J. 10, 17, 32, 35
Nguyer, H. 10, 16

Patrick, D. L. 173
Pauker, S. G. 116, 148
Pearce, D. W.
Perlman, M. 35
Pliskin, J. S. 115, 120, 148, 173

Quade, E. S. 76, 111

Raiffa, H. 173
Read, J. L. 174
Reynell, M. C. 8, 9, 17
Reynell, P. C. 8, 9, 17
Richardson, A. W. 42, 67, 171
Robinson, G. C. 32, 34, 78, 111
Russell, I. T. 10, 17, 32, 35

Sackett, D. L. 6, 11, 17, 27, 34, 35,
    45, 67, 94, 111, 115, 116, 117,
    118, 119, 128, 129, 134, 148,
    173
Schoenbaum, S. C. 174
Shepard, D. S. 115, 148, 171
Shoemaker, P. J. H. 116, 147
Sinclair, A. J. 117, 148
Sinclair, J. C. 14, 16, 27, 34, 35, 42,
    66, 119, 139, 147, 148, 171
Stason, W. B. 14, 17, 28, 29, 35, 81,
    111, 113, 148, 161, 162, 171,
    172, 173
Steele, A. 1, 4, 175
Stein, L. I. 13, 17, 19, 35, 156
Stoddart, G. L. 2, 4, 11, 17, 21, 27,
    34, 35, 45, 67, 78, 94, 111
Struening, E. L. 10, 16, 67, 171, 172
Sugden, R. 150, 152, 167, 174
Sullivan, B. 2, 4
Sutherland, H. J. 116, 119, 120, 147,
    148

Tessier, L. 10, 16
Test, M. A. 13, 17, 19, 35, 156

178

Subject index

Subject index

cost (*cont.*)
  common, *see* cost, shared
  direct 21, 22, 142
  equivalent annual 42, 51, 68
  fixed 43
  hotel 45, 90
  indirect 21, 22, 48, 78, 142, 157, 164
  incremental 30, 63, 98, 112, 160
  intangible 2, 152
  joint 158
  marginal 43, 65
  opportunity 7, 41, 42, 66, 108, 151, 152
  overhead 21, 25–6, 39, 43, 45–7, 53–6
  psychic 21–2, 153–5
  shared 26, 46, 96; *see also* overhead cost
  treatment 90, 153, 157, 164, 166
  total 43, 45
  variable 21, 43
cost function 43
cost–analysis 2, 4, 8, 9, 39–73, 170–1; *see also* differential timing and overhead costs
cost–benefit analysis 1, 3, 11–15, 144, 146, 149–67, 174–5; *see also* costs and benefits
cost–benefit ratio, *see* benefit–cost ratio
cost–effectiveness analysis 3, 4, 10–11, 15, 74–111, 144, 151, 154, 162; *see also* costs and effects
cost–effectiveness ratio 31, 37, 98, 106, 109, 112, 146, 163, 167, 171–2
cost-minimization analysis 3, 10, 15, 62, 75
cost–utility analysis 1, 3, 14, 15, 28, 64, 75, 112–47, 151, 154, 156, 173–9; *see also* utility
costs and benefits
  direct 2
  discounting of 81
  indirect 2, 78–9, 96, 157, 164
  intangible 2, 149, 152–3
  psychic 153–5
costs and consequences 2, 3, 7, 8, 9, 19–25, 29, 30, 36–7

  identification of 21–5, 36, 62–3, 96, 106, 142, 157–8, 163–4
  measurement of 25–6, 36, 63, 96, 98, 106–7, 142–3, 158–9, 164–5
  qualitative treatment of 165–6
  valuation of 26–8, 36–7, 62, 97, 108, 143–4, 149–50
costs and effects 143
  discounting of 29, 166
  incremental analysis of 166

day care surgery 32
decision-making approach 152
decision-making under uncertainty, *see* uncertainty
decision trees 87–9, 172
depreciation 42, 66–9
differential timing 29, 37, 39, 48–53, 64, 97, 109, 144–5, 160
direct cost, *see* cost
discounting 29, 39, 48, 80–1, 134, 138, 144, 160
discount rates 51, 66, 82, 144, 145, 167
discount tables 71–3
distributive justice 33, 99, 150, 152–3
double counting 160

earnings of patients, *see* patients' earnings
economic efficiency 9, 18, 31, 141, 145, 150, 152, 158; *see also* welfare economics theory
economic evaluation 5–17
  assessment of 18–38
  components 2
  definition of 7, 8
  elements of 18–32
  full 8, 9
  importance of 6–7
  limitations of 16, 33
  partial 8, 9
effects 2, 13, 23, 24, 61, 162
  discounting 81
  multiple 11
  natural units of 15, 30
  single 11
  see also costs and effects

Subject index

The authors are grateful to Anne Mason and Sue Elias for help in compiling the index.